Roger French is University Lecturer and
Director of the Wellcome Unit for the
History of Medicine, Cambridge.

Frank Greenaway is a Reader in History
of Science at the Royal Institution,
London.

SCIENCE IN THE EARLY ROMAN EMPIRE:
PLINY THE ELDER, HIS SOURCES AND INFLUENCE

Science in the Early Roman Empire: Pliny the Elder, his Sources and Influence

Edited by
ROGER FRENCH
and
FRANK GREENAWAY

BARNES & NOBLE BOOKS
Totowa, New Jersey

©1986 R.K. French, Frank Greenaway
First published in the USA 1986 by
Barnes and Noble Books
81 Adams Drive
Totowa, New Jersey, 07512
Printed in Great Britain

Library of Congress Cataloging-in-Publication Data

Science in the early Roman empire.

Includes index.
1. Pliny, the elder—knowledge—science. 2. Science
—Europe—history. 3. Science, ancient. I. French, R.K.
(Roger Kenneth) II. Greenaway, Frank.

Q143.P64S35   1986   509.37   86-7919
ISBN 0-389-20634-2

CONTENTS

ACKNOWLEDGEMENTS

The Symposium of which this is a record could not
have been held without the encouragement and co-
operation in a variety of ways of several bodies
in addition to the Royal Institution which, through
its Centre for the History of Science and Technology, was the prime mover. Thanks are due to
the International Academy of the History of Science.
the International Union of the History and Philosophy of Science, the Society of Antiquaries of
London, the British Society for the History of
Science, the Society for the History of Alchemy
and Chemistry, the Wellcome Institute for
the History of Medicine.

Typographus ad lectorem

Ut Plinius
ita typographus
magnis laboribus pluribusque
erroribus opera aliorum varia transtulit
Candide
remitte hanc noxam
ambobus

INTRODUCTION

This volume started life from a thought less concise than its title: that while there is continuing interest on the part of many historians in the scientific thought of early Greece, in the development of Islamic science, in Chinese science and even in pre-Columbian science, little attention is paid to Rome, geographically or culturally perceived as a feature on the scientific landscape.

There may be good reasons for this. The philosophers of recent times find their sources for debate in earlier writers in Greek. Roman thought appears so often derivative. Almost the only Latin writer pointed out to the tyro student in the history of science is Lucretius, and the alert student so soon finds that to grasp his scope one has to enjoy his poetry that it is all too easy to bundle him up with Epicurus and the atom and leave it at that. Such Greek influences on Roman thought have been so well rehearsed that the scholar trying to comprehend the liaison between the ancient classical science and the modern may well be excused for thinking of the Roman achievement as so much technological triumph followed by decline and fall. This is not enough.

During discussions at the Royal Institution Centre for the History of Science and Technology, these questions of the continuity of science led us to propose a meeting in which scholars interested in some aspects of the science of the period should be able to talk to each other and offer some report on their current work in or around the edges of the field. As ever, the main value of such a meeting was personal, and we believe that we may count on improved links between those who attended to talk and to listen. However, the present volume may well bring others into the discussion.

One could well take up a great deal of time in

considering whether there was anything that could be called ´science in any modern sense in the long centuries of Roman domination. One could also take time in considering what might be identified as Roman and what as Hellenistic, but this would not further the search for what was scientific.

However, does it matter if we identify the scientific´? We think so because we do not yet know what that means in our own time. There is work in progress on the recognition of a scientific faculty in material culture at large, which may well change some of our views of the character of the activities which led to the emergence of the forms of civilization. It is of equal importance that we should identify the ´scientific´ at times with which we do not usually associate it.

These essays therefore embody a wide variety of argument and evidence, some reporting recent research, some constructively speculating, some of it deliberately provoking so as to prompt further enquiry.

One might say all that about Pliny the Elder and his great **Historia Naturalis,** which seemed to us the reference point to which we should refer our contributors. New translations are being prepared in Great Britain and Russia, new light is being thrown on his methods of work. There is an interplay between his methods and his social role: he was able to indulge his interest in the natural world because he travelled widely as an effective official, and he was all the more assiduous as an official because he was kept constantly alert to the world in which people had to live and make the decisions which made survival possible. This is our problem too, and so it is to be hoped that these pages may reflect Pliny´s concern for the common good.

Frank Greenaway

Chapter One

THE ELDER PLINY AND HIS TIMES

J Reynolds

Pliny(1) came from Transpadane Gaul, no doubt from Novum Comum, where the connections of his nephew and adopted son, the younger Pliny, are well attested. He was born in late AD 23 or early in 24, since he was in his fifty-sixth year when he died in August 79. No certain record of his parents has yet been found. Comum was a town whose population was ethnically mixed, since it had twice received colonists in the first century BC, but the father's name could indicate a family of native stock; whether it did or not, it is certainly more important that it was a family of wealth and standing, for Pliny's sister married the owner of considerable property in the area and he himself was able to embark on a Roman career open only to men of Equestrian status, that is, of the second order in the Roman social hierarchy. The family, we may take it, belonged to the municipal governing class; and that is already information of use to us. For men of municipal origin, so the historian Tacitus observed, writing a generation later than Pliny, tended to favour stricter and more antique codes of behaviour than those characteristic of the imperial court and the Roman aristocracy in the middle years of the first century AD, and their attitudes commonly survived transplantation from municipal homes to Rome. The younger Pliny makes something of the same point, with special reference to Transpadanes, though he is less trenchant (2). We have then a natural explanation of the elder Pliny's well known and persistent inclination to castigate what he regarded as luxury, and to see in its lures an adequate explanation for features of contemporary life which he deplored. He represents in this attitude a recognisable social phenomenon of the period: that of men brought up in a comparatively

1

simple life style and reacting with somewhat puritanical naiveté to a more sophisticated one. It is not very easy to translate these theoretical terms into realities, for we have all too little precise evidence of the material trappings of everyday life either in the municipalities or in the city of Rome, apart from imperial palaces. Pompeii is one of the few sites from which such evidence is available and, although it is far from certain that it was typical of municipalities, it is worth drawing upon its testimony. The House of Menander, the major house under survey by the British team now working at Pompeii,(3) has many fine features of the first century AD date - notably its large peristyle garden, comfortable dining room and bath suite, its attractive wall paintings, which include a representation of the playwright Menander that gives it its modern name, and a splendid equipment of silver vessels, each piece carefully wrapped in cloth. There is no ostentatious excess, but there is certainly comfort as it was understood at the time; and some degree of elegance. At the same time, in the back quarters, not visible from the grand apartments and the garden, and not, it is suggested, to be smelt from them either, there were, in August AD 79, a stable, a farm cart, stores of farm tools and produce (or containers for it); apparently the equipment of a working farm, although its fields must have lain a little distance away, outside the city walls, one presumes. So this was not just a town house, but a farmhouse too in a sense; and I doubt if an elite town house in Rome could have doubled its function in this way. Further, there are signs that, also in the first century AD, parts of the premises were adapted for profitable exploitation; ground floor rooms with street frontages were converted into shops, and in the next-door House of the Lovers an upper storey was built which seems to have been let to a brothel keeper. It is not of course in doubt that property owners in Rome let apartments for residence and shops; but it would be surprising, perhaps, if the great aristocrats shared their town houses with tenants like this as a regular thing. Among additional differences one can point to the relative infrequency in Pompeii of expensive materials like coloured marbles and imported woods, and the relatively limited space for domestic staff, by contrast with what the literary sources suggest for the great and luxurious at Rome.(4) So we have some quantitative and qualitative clues, if not enough, to what shocked the municipal men.

But whatever the differences in life styles, many municipal men were, none the less, keen that their sons should play a part in the Roman social and public world, Pliny's parents among them. They gave him not only the necessary property qualification, but also the 'gentleman's education' essential for a start, and they must have introduced him to the city of Rome at an early age, since he writes, from time to time, as an eye witness of events occurring there in the thirties of the first century AD.(5) So his municipal experience was broadened under metropolitan influences. Although his home grown distaste for luxury survived, it is clear that he reacted with interest to very much of what he found, including both old and new in literary movements, theatre, art, oratory, philosophy, science. This volume is intended to treat of his transmission of scientific matters; but in attempting to assess that it is pertinent to remember that he did not specialise in any one subject - he was a man reasonably informed in many, if not very sensitive to their nuances. He also practised for a time as a pleader, must have been constantly concerned with the running of his estates, and managed to fit in a public career involving service as an army officer and several posts in public administration as well.

It was, in his time, a natural and a proper move for a young man of his social status, to seek at least one posting as an army officer. The emperor Augustus had in fact encouraged a view that to hold public status as an Equestrian involved something like an obligation for a young man to do so. Not all accepted an obligation; and certainly not all who accepted a posting held it for long or put great effort into it. Many would be no more than 'pen soldiers' anyway, stationed in peaceful areas, occupied mildly with exercises on the parade ground, a little police work, supervision, perhaps, of road or canal construction, expected to cast an eye on the records and accounts kept by the regimental clerks, but free often to indulge in reading (if they had books), writing (if that was to their taste), hunting, sightseeing or whatever else seemed to them conducive of the good life (and Pliny has some tart remarks on the grand dining service taken to his station by one officer of his acquaintance).(6) But Pliny in fact served in several posts, which suggests a quite serious interest in this part of his career; and served them

in Germany, where real military action might occur.
It is reasonable to deduce from his reference to
autopsy in Germany that he actually took part in a
campaign east of the Rhine conducted by Domitius
Corbulo, governor of Upper (north) Germany in AD 47
and that, at a later date, he was posted also to
Lower (south) Germany, probably during the
governorship of Pomponius Secundus in AD 50/51.
Still later, he served another tour of duty, likely
to have been in Germany too, when his fellow
officers included Titus, son of a later emperor
Vespasian (7), to whom, in due course, he dedicated
the **Natural History.** We are told that in one of
the three posts he commanded an auxiliary cavalry
regiment (**ala**); almost certainly not the first, to
judge by what we know of other Equestrian military
careers of the first century AD. During the reign
of the emperor Claudius the posts available were
roughly grouped in such a way that commands of
auxiliary infantry regiments (cohorts) were the most
junior ones, commands of auxiliary cavalry regiments
on the next rung up the ladder, and officerships in
legions (as military tribunes) above them; but since
this order did not correspond well with the actual
responsibilities attached to the second and third
rank posts, these changed places after Claudius´
death; but we do not know whether it was soon enough
to affect Pliny. It is highly probable that his
first post was command of an auxiliary infantry
regiment; but at this date there was no strict
rigidity of career structure, and a variety of
combinations are attested - a man might hold several
posts in one rank before moving to another in a
higher ·rank, so that there is no saying for certain
whether Pliny held his cavalry command as the second
or third of his posts, although I should guess that
it was his second.

To get any posting it was necessary to have
patronage. Pliny wrote a two-volume memorial
biography of the senator Pomponius Secundus, under
who he had probably served in AD 50/51 and whom he
had certainly known before he took up any posting;
and he is said to have done it because he felt he
owed it to so good a friend.(8) ·That sounds as if
patronage by Secundus had been a factor in his
career; and it adds yet another point to our
knowledge of his character - that he was a man who
took a debt of gratitude very seriously. He would
of course have been a satisfactory candidate to back
for postings - to write testimonials for or to
appoint to one´s own staff - since he also took his

duties seriously. For this the evidence is not simply that the younger Pliny describes him as serving **industrie** (that is hardly unbiased testimony); but the record of his early writings. His first publication was an essay on cavalry manoevre, the **De Iaculatione Equestri,** which certainly demonstrates that he was taking an interest in what he was supposed to be doing. It is also evident from the **Natural History** that he kept his eyes and ears open, curiosity alert, to see what he could see and learn in or near the frontier area (9). Moreover, his third literary undertaking (after the biography of Secundus), apparently begun in Germany, or at least projected there, though probably not finished until later, also shows how his attention was caught by his current location and concerns; that is what became the twenty-volume **History of Rome's German Wars.** For this, he is said to have explained, the impulse came from a dream (10); which may seem to us childish, but should not be treated too dismissively. It is fully in accord with the ethos of his times (11), and while it leaves us clear that his intellectual powers were not such as to raise him above the run of his contemporaries in this matter, it is not evidence for the kind of weakness that we might associate with it today. The details of this dream are themselves revealing; he imagined that he saw the dead Drusus, a stepson of Augustus and a general in important and successful campaigns in Germany, who asked him to rescue his memory from oblivion. It is true that in the surviving historical tradition, successes in Germany won after Drusus' death, by his brother, the later emperor Tiberius and to some extent by his son Germanicus, have attracted more attention than his own; and it may be that in the frontier areas there was more to remind an enquiring mind of the part that Drusus had played. But Drusus was father to the emperor Claudius, who himself did something to assert that part (12); and Pliny is probably responding, no doubt without complete consciousness of it, to an imperially-set trend. It is useful to have this early indication of his susceptibility to current fashion, his willingness to accept something of the attitudes and the formulae of a courtier. The same susceptibility can be traced to later reigns; and it is of course in the dedication to the **Natural History** to Titus that the courtier's approach is the most explicit. Here, in the **History of the German Wars,** at least three strands in Pliny's thinking are

intertwined - his youthful interest in military
affairs, his curiosity about the German frontier on
which he served and his response to a lead given by
an emperor.

After the books on the German wars, his
bibliography (13) shows a change of direction, which
must be associated with his final return to civilian
life. Here belong three books on the development
of an orator, which presumably coincided with and
grew out of his activity as a pleader. His private
financial concerns must have occupied him too, but
of that there is only a very uncertain hint: a
passage in the **Natural History** refers to a
particularly good variety of cherry grown in
Campania and called **Pliniana**(14), which sounds as
if someone with his name had been successfully
introducing a new variety of fruit tree. Such
activity was in the tradition of Roman estate owners
- Pliny refers to several contemporary examples (15)
- so if he was himself responsible, he would again
have been following a fashion; but in fact he makes
no personal claim here, so may simply be recording.
In addition he kept up to date with what was going
on in Rome and sought to follow current artistic
developments; in the wake, no doubt, of another
imperially-set fashion, for the young Nero, who
succeeded Claudius in AD 54, was of course
remarkable for his artistic enthusiasms and talents.
Subsequently Pliny expressed himself distinctly
critical of Nero (16); but Nero's reign ended in
disaster and a series of civil wars, while Pliny's
career blossomed in the new stability created by the
subsequent Flavian dynasty under Vespasian and his
son Titus. Nero's crimes were a feature,
naturally, of Flavian propaganda for their own
regime, so that it is not surprising that Pliny
wrote as he did about him; we do not have to assume
that he had personally suffered from Nero; it could
of course be so, but, if so, we might have expected
that the younger Pliny would have made an explicit
point of it. So it may well be that, at the time,
Pliny was satisfied with the life of a writer and
landowner, who also occasionally pled cases in the
courts, dividing his energies happily between Rome
and his estates, taking an interest in intellectual
developments but not anxious to bring himself to the
attention of politicians. At any rate he visited
the Roman **atelier** of Zenodorus, who had been
commissioned to make a colossal bronze statue of
Nero, and recorded something of his techniques;(17)
he knew about and perhaps watched the artist at work

on the wall paintings that were a notable feature of
the new palace that Nero was building, the Golden
House. Later he remarked acidly that since his
talent was expended largely on those walls, his art
was, for that reason, a prisoner (18), but it does
not follow that he thought in those terms from the
first. He was also beavering away, making records,
both from his reading and from personal
observations, of anything and everything that seemed
to him worth knowing. He had begun earlier - that
is clear from the snippets of information that he
introduces in the **Natural History** from the Rome of
his youth and the Germany of his military service.
There was no slackening of the effort now. It was
an effort that reflected what seems to have been an
enormous contemporary curiosity, an appetite for
knowledge, not very discriminating, but very real,
often drawn to marvels, whether natural or man made,
but not confined to them. That many were as
assiduous as Pliny in collecting and recording is
most improbable, but quite a number did it in a most
desultory way. Pliny is, again, illustrative of a
feature of his age. It can be seen at one level in
Nero´s excitement over a water-organ, at another in
the publication (perhaps a travel journal) of the
senator Licinius Mucianus (and he was one of the
most powerful figures in the coup that brought the
Flavian dynasty to the throne) and at yet another in
the very considerable sum of money offered to Pliny
in Spain by a senator who wanted to buy his
notebooks.(19) As noted above, the information
collected and retailed was not all second hand;
autopsy played a part; but there do not seem to have
been any planned programmes of observation of
particular categories of phenomena and still less
anything designed to produce explanations. Facts
were enough; apparently for everyone, certainly for
Pliny.
    We do not know, unfortunately, what Pliny was
doing in the troubled months that ended Nero´s
reign. The younger Pliny implies that he was in
Italy, engaged in the unobtrusive study of language,
on which he wrote eight books; he passes over the
following period of civil war without a word,
presumably because Pliny continued to avoid limelight
until the accession of Vespasian was assured in 70,
when he might be able to benefit from the patronage
of his former fellow officer, Titus, the new
emperor´s elder son. Thereafter, and before his
death in August 79, he wrote thirty books of a
History of Rome and the whole of the **Natural**

History, while holding several administrative posts of Equestrian rank and reasonably high seniority (the biography attributed to Suetonius suggests at least three), as well as becoming an official friend of the emperor, which meant that when in Rome, he would be expected to wait upon the emperor every morning, and might be called upon to attend business sessions as an advisor. The posts he held certainly included the procuratorship of the largest and most important division of Spain, Hispania Tarraconensis, where his responsibility will have been essentially for the emperor's financial affairs, and he was probably there early in the reign when a census was held, yielding information of which he drew;(20) also the prefecture of the fleet at Misenum, an administrative rather than a military function, which he held at the time of his death. Others have been conjectured. On the basis of the personal observations that he records, we may be sure he had visited Africa and Gallia Narbonensis (Provence).(21) He might perhaps have been to Narbonensis privately, conceivably diverging on the way to or from one of his postings in Germany; but North Africa is off any likely private track, and it is much more probable that he went there on official business, as a procurator. It is usually thought that a post of this type was held for at least three years, and if Pliny's postings were all of that length there is not so much time left for more than these three. But there was no rule about these things - the emperor could do as he pleased; Vespasian might argue that a conscientious and sensible man could be most useful in the aftermath of civil war if he received a series of short term appointments to each of which he could bring his particular qualities with advantage. So space could be found for a posting to Narbonensis, at a pinch. But whether or not he did hold four posts in this period, there is no doubt that Pliny was a very hard worker; all his life, but especially in this last decade; and for this too there are parallels to show that he was no original, above all in that last decade the parallel of the new emperor Vespasian.(22)

So we end with a man who exemplified important features in the society of his time, who played a minor part in the public affairs of the Julio-Claudian period and a rather larger one under Vespasian, his career a product of the system of patronage which governed Roman life; hard working, moralistic; open to the interest of the intellectual trends of his day and accepting many of its fashions, though not all; seeing nothing wrong in setting out to please an emperor (but there is no evidence that he did so against his conscience); above all a glutton for information, curious about, though not seriously questioning about, all that came his way; constantly reading or being read to; wasting no minute; believing that even the worst book might yield a grain of fact; but not a man with his eyes wholly glued to the written word. In the end it was curiosity to see - combined, it must be said, with humanity - which took him to his death, when he put out from the fleet base at Misenum on the 24th August, 79, after his sister had drawn his attention to the odd cloud above the mountains, stone-pine shaped, or, in our terms, mushroom shaped, which he thought worth further observation (though urgency was added by the call of a friend for rescue from the sea). While sailing to observe and rescue, he took, we are told, continuous notes. Later he found retreat impossible, and died on the shore.(23) The younger Pliny does not say any more of these notes in his account of his uncle´s experiences of the eruption of Vesuvius and of the earth tremors that accompanied it; but it makes a fitting end to his uncle´s life that, to within the last hours, he was still recording the personally observed marvels of nature.

NOTES

1. A brief account of the life and times of the elder Pliny seemed a useful preliminary to the serious business of the colloquium. What is offered here contains nothing new; but is intended solely to invoke something of the character, attitudes and experiences that lay behind his work. The organisers of the colloquium had hoped that new light might be thrown on Pliny by the work of the Pompeii Research Committee, being carried out under the direction of Dr Roger Ling, with funds provided by the Imperial Tobacco Company. This, however, is essentially the recording of a previously excavated, but inadequately published **insula.** It is revealing social and economic developments from the fourth century BC to AD 79, and, for the period of Pliny´s life, the results provide background evidence which may be useful to our understanding of him - but at present it takes us no further.

Most of the ancient evidence for Pliny's life is contained in the letters of the younger Pliny, **Epp.** III.5 and VI.16,20 (for commentary see A N Sherwin-White, **The Letters of Pliny,** 1966); there is also a short biography attributed in the manuscripts to Suetonius; more can be deduced from the **Natural History.** The most recent and authoritative account of his career is by Sir Ronald Syme in **Harvard Studies in Classical Philology, 73** (1969), pp. 201ff.

On Transpadane Gaul see G E F Chilver, **Cisalpine Gaul,**1941, especially pp. 106-7, and on Novum Comum the short entry in the **Oxford Classical Dictionary,** 1970 s.v.

2. Tacitus, **Annales** XVI.5; Pliny, **Epp.**1.14.4
3. For accounts see R Ling, **Current Archaeology,** 85,(Dec. 1982) pp. 55ff.
4. Vitruvius VI.5 is useful here.
5. Thus **NH** IX.117, XII.10, XIV.56.
6. **NH** XXXIII.143
7. So Syme, loc. cit. in n.1.
8. Pliny, **Epp.**III.5.3; cf. **NH** XIV.56
9. Thus **NH** XVI.2, XXXI.20.
10. Pliny, **Epp.** III.5.4.
11. Cf. Sherwin-White, **Letters,** in n.1,p.128 on Pliny, **Epp.**I.18.1
12. H Dessau, **Inscriptiones Latinae Selectae,** no. 208
13. Pliny, **Epp.** III.5.5ff.
14. **NH** XV.103.
15. Thus **NH** XV.47.
16. **NH** VII.45,46.
17. **NH** XXXIV.45ff.
18. **NH** XXXV.120.
19. Suetonius, **Nero,** 41; **NH** VII.36, VIII.6,for instance.
20. **NH** III.28
21. Thus **NH** II.150, XVII.41.
22. Suetonius, **Vespasian** 21.
23. I had accepted an old convention that he died of asphyxiation; for a re-examination of the evidence and a rejection of this convention see Conway Zirkle, "The death of Gaius Plinius Secundus", **Isis, 58** (1967) pp. 553ff., and L Bessone, **Rivista di Studi Classici, 17** (1969) pp. 166ff.

Chapter Two

THE PLINY TRANSLATION GROUP OF GERMANY

R C A Rottländer

Growth of knowledge is very often governed merely by chance, and this account of the German Pliny Translation Group is a new chapter of this old story.

At a meeting in June 1976 of the German Archaeometry Group, professor Schaaber read a paper on a re-examination of some of Pliny's remarks on metals. In passing he mentioned that he had observed that Pliny used a sort of key-word system. Schaaber (who was working with modern translations) called these key-words **descriptores** and explained the way in which they had been used.

His remarks startled me, for I had made a similar observation in Pliny's chapter on amber. Together with a few others, we decided at that meeting to establish a Pliny Translation Group. The aims of the group were two.

First, to undertake a new translation of parts of the **Natural History** by means of cooperation of natural scientists, philologists and historians, and aiming to produce a clear modern German version for the non-specialist.

Second, to look for further proof of the system of **descriptores** with a view to gaining a deeper insight into the internal structure of the **Natural History.**

The first meeting of the new group, chaired by professor Schaaber, was in 1977, and its purpose was to begin with Pliny's chapters on glass. To this end the group (about a dozen in all) contained a specialist in glass manufacture, chemists and physicists as well as historians and a philologist. After several meetings (there are normally two or three a year) it became clear that the following technique was the most effective: first, a ´raw´ translation is provided by the philologists (two at present) and distributed in typescript to the remainder of the group (about 20 at December 1983). Comments are gathered together by the chairman and circulated round the group some time before the translation session itself begins. At the session each sentence of the ´raw´ translation is exposed to the comments of the various specialists, and the implications are considered by the group as a whole. If there is no agreement about the final translation, then the more probable interpretation is adopted and the less probable given as a footnote. In case of ambiguity or obscurity in the Latin, the relevant scientists are asked, where appropriate, to illuminate the text by conducting laboratory experiments.

There is a threefold apparatus to the text. First (indicated by Arabic numerals) are noted textual obscurities of the kind mentioned above, together with problems of interpretation. For example, sometimes a Latin word cannot be translated into a single German word, just as the Latin **aes** may in English indicate copper, brass, bronze or pewter.

Second (with Roman numerals) is noted what light can be thrown on a topic by science and technology. This involves not only a consideration of the history of the topic, but also the results of the experiments referred to above. Lastly (marked with Roman letters) are listed references to the same topic in other ancient literature, together with any further relevant writings.

In this way the original ´raw´ translation is refined; often it is necessary to carry the process over a number of meetings, when further experimental results and any additional criticisms of the translation are added, but this can normally be a relatively quick process (although again interim versions of the translation are circulated between meetings). To date the group has concerned itself with glass, iron and **aes**; of these, the translations relating to glass and iron have been published, and the first part of that concerned with **aes**, that is, its technical, metallurgical aspects, is in press (at December 1983). The current concern of the group is the medical section of the **Natural History** which, it is hoped, will be completed during 1984. Topics for future years include gold, silver, tin, lead, pigments and finally amber.

The strength of the group lies in its combination of human and natural science; it will perhaps be appropriate to give an example of how that combination has corrected long standing errors of translation.   The Latin text of book 36, chapter 194, reads (shortened for convenience):

Iam vero et in Volturno...harena alba nascens...qua molissima est, pila molave teritur.   dein miscetur tres partibus nitri pondere vel mensura...

    The ´raw´ translation of this was close in sense to that of Eichholz in the Loeb edition: "Then it is mixed with three parts of soda...".   But is was at once clear to our glass scientist that such a mixture would never become glass, which is mainly silica, because three parts of nitre or soda would provide far too much alkali.   The mixture would result in a water-soluble mass of no use in the construction of vessels.   The group naturally had to face the problem of whether Pliny had mistaken the procedure or the text had become corrupt. Making the decision that Pliny had reported the technique correctly, the problem for the group became a textual problem, and was referred to the philologist.   His solution was that **pars** can be translated as ´part´ as well as ´quarter´, as other Latin texts demonstrate, and the meaning in each case is to be determined by the context.   And this particular sentence has been recognised as open to interpretation since the time of Blümner at the end of the last century.
    Another example may be taken from where Pliny is discussing iron (book 34, chapter 143).   Vetters argues convincingly that Pliny´s **brevitate   sola placet   clavisque   caligariis...** should not be translated ´another   variety of iron finds favour in short lengths only and in nails for soldiers´ boots...´, but rather ´one (sort of it) is particularly suited due to its low dimensions for hobnails...´.   That is, these small lumps of iron would present an advantage in the manufacturing process.
    At the end of this chapter Pliny has **stricturae vocantur hae omnes, quod non in aliis metallis, a stringenda acie vocabulo inposito...** that is, in referring to all the foregoing kinds of iron as being capable of being drawn out into a cutting edge.   Certainly **stringere   aciem** must have originally have meant simply ´to draw out a sharp edge´, as Rackham translates; but Schaaber, a specialist in the hardening of iron to form steel, argues that it had by Pliny´s time become a technical term, meaning ´forming a point of steel´. Such a distinction, between iron and steel,

is perhaps also supported by archaeological evidence: Vetters points out that there are known, from the Latène period and in Roman times, bipyramidal ingots of iron, of which one of the two points has been left as soft iron, as can be shown by bending it, while the other in contrast has been hardened with carbon to provide a steel capable of being reworked with heat. It may be that the French word for steel, **acier,** reflects the **acies** of the phrase above; but it cannot be proved.

Similarly, where Pliny continues his description of different kinds of iron with **nucleusque quidam excoquitur** (translated by Rackham as "a certain knurr"), Vetters and Schaaber conclude that **nucleus** is another technical term for iron with a high carbon content, suitable for forging cutting edges. Further, Pliny continues his account of the varieties of iron with **secunda Parthico. neque alia genera ferri ex mera acie temperantur, ceteris enim admiscetur mollior complexus...** This is translated by Rackham as "The second prize goes to Parthian iron; and indeed no other kinds of iron are forged from pure metal, as all the rest have a softer alloy welded with them." Again, we take acies as ´steel´ and not for pure iron, firstly because ancient ideas about what is a pure metal are quite different from ours. It seems indeed that their names for metals depended on the place of origin of the metal - as in this chapter of Pliny´s - rather than on any notion of ´purity´. Secondly, this new meaning of **acies** is supported by frequent archaelogical finds in which the cutting edge of implements is steel and the remainder soft iron. And a third reason for disagreeing with Rackham´s translation is the improbability of the Romans having formed an idea of an alloy in the modern sense. As far as we can see they believed for instance pure copper in our sense to be an impure metal which had to be washed to get it more pure. They considered the addition of tin to be a proper measure to make the copper cleaner, because its ·colour increasingly approached - but never reached - the colour of gold. More effective, in ancient opinion, was the addition of **cadmea,** calamine in our terms, because the result was an even more goldlike metal: our brass. This is reflected in the Roman system of coins: the **as** is of pure copper, the **sestertius** of bronze, but not four times as heavy as the **as;** while the **dupondius,** though of similar size to the **as,** had twice the value and could only be distinguished by the yellow colour of the

14

aurichalcum. With this in mind, our translation of the passage is: "The second rank belongs to Parthian iron, because nowhere else are iron implements manufactured totally from steel; elsewhere the steel is embedded into soft iron". Obviously these examples represent not so much the correction of errors of translation as illumination thrown on a difficult point by scientists and specialists.

Another matter in which we hope to have eliminated an old error is the question of the identity of plumbum argenterium. Rackham gives a note to the effect that it is "silver lead" consisting of half silver, half lead. This seems in fact not to be the case; and the evidence is presented below. First, as to plumbum, all authors are agreed that plumbum album is tin and that plumbum nigrum is lead; and we also agree. It is stated by S Boucher, who investigated the nature of of plumbum argenterium, that in a recipe described by Pliny (XXXIV.98) plumbum nigrum and plumbum argenterium are both added to aes, and so must be two different things. We accept this, but are unable to agree, for the reasons given above, with Boucher´s further statement that aes in a recipe must always be pure copper. However, Boucher found that tin is never added in a recipe for bronze, and she therefore concludes that plumbum argenterium must be tin. But this contradicts the fact that tin is plumbum album. On the other hand, we know that bronze was produced by adding tin ores, mainly cassiterite, during the process of smelting. The idea behind this was to wash the copper, for the notion that the weight of the tin-stone would increase the weight of the bronze was not yet born.

As proof of her opinion, Boucher refers to Pliny´s recipe for the production of statue bronzes. The mixture is prepared as follows: to a given amount of molten aes one third of scrap aes is added, and for each 100lbs of the melt, twelve and a half pounds of plumbum argenterium are added. Boucher believes that analysis will reveal that tin is meant. Yet according to analyses made by Riederer in Berlin the tin content of statue bronzes is typically under ten percent, whereas lead is typically over ten percent, sometimes surpassing twenty. Of course, the scrap metal which is added brought with it its own tin and lead. So plumbum argenterium must be lead and not tin. But what then is the difference between plumbum nigrum and plumbum argenterium? To answer this question we have to look at the old way of producing silver.

Silver ores never contain a high proportion of silver; and silver is found also in certain lead ores, for example those of Laurion. In order to obtain the silver, it is first necessary to separate the lead, with which the silver forms an alloy. In a separate open air process the lead is first oxidized to form litharge, leaving the silver as an unoxidized residue. This litharge is very valuable, because it is pure lead oxide and lead can be recovered from it very easily by reduction with charcoal. Lead produced in this way is very clean and, not very inclined to oxidize in air, remains bright for a long time. In contrast normal lead contains many traces of other elements and is for this reason more easily oxidized, becoming dark in colour: **plumbum nigrum.** ´Bright´ lead, then, was a by-product of the silver industry, **argentaria,** and there are extant a large number of ingots, particularly from England, bearing the inscription **ex argent.** It is not inconsistent with this explanation that lead ingots with a relatively high silver content are found: under the given circumstances it was probably not worth while to extract the silver, or possibly the ancient analysis of the lead ore was erroneous.

Yet another error of translation, we are convinced, is in chapter 96 where Pliny wrote ...**et carbone recocunt propter inopiam ligni....** This is normally translated as Rackham translates, "...and do additional smelting with charcoal because of their shortage of wood." It is not clear how a region which is short of wood could be sufficiently supplied with charcoal, which is produced from wood. If fuel had to be imported then presumably wood could be obtained as readily as charcoal. It seems very much more likely that **inopia ligni** is not ´shortage´ of wood, but rather ´inability´, since a wood fire does not generate enough heat to melt large pieces of copper.

A related problem occurs at the end of the same chapter, where Pliny has ...**aes omne frigore melius fundi...,** which is easier to translate than to understand. Rackham renders it "...all copper and bronze fuses better in very cold weather". Our metallurgist tells us that the results of casting are better in the cold because the crystals formed are smaller and the metal less brittle; so what Pliny seems to be referring to is simply the first part of the process, the melting of the copper or copper ore: the means of production but not the end.

Another case of difficulty in understanding what

Pliny has to say occurs in chapter 112, where Pliny is dealing with the use of a hot shovel in testing verdigris for adulteration with shoemaker´s black. Pliny´s words are ...**experimentum in vatillo ferreo, nam quae sincere est, suum colorem retinet, quae mixta atramento, rubescit...**, rendered by Rackham as "...as a specimen that is pure keeps its colour, but what is mixed with shoemaker´s black turns red", to which he adds a note that it is not true. A lengthy discussion on this point resulted in our deciding that the Latin allows us to interpret **suum** in a slightly different way, that is, to imply that the verdigris has its proper colour in the circumstances, black. This is confirmed by experiments in the laboratory. Further, we tested copper salts (thought to be in shoemaker´s black: see Rackham) with gallic acid. No changes were observed over two days, and so we are convinced that shoemaker´s black (**atramentum sutorium**) can only be a ferric, not copper, salt, since water-soluble ferric salts immediately give a black colour when brought together with gallic acid, which is always present in leather from the manufacturing process. A few lines further on, Pliny has ...**deprehenditur et papyro galla prius macerato, nigrescit enim statim aerugine inlata** ("It is also detected by means of papyrus previously steeped in an infusion of plantgall, as this when smeared with genuine verdigris at once turns black." On the basis of the chemical reactions just mentioned it will be seen that to translate **aerugo** as "genuine verdigris", as Rackham and others do, must be an error. Pliny was here concerned with adulterated verdigris and was listing all possibilities of detecting forgeries. The addition of shoemaker´s black does indeed cause the formation of a black colour; so clearly Pliny intends adulterated verdigris.

As a last example of laboratory experimentation, I want to discuss a passage in chapter 95, about the colouring of **aes**. The text has come down to us in two versions. The first is ...**ita lentescit coloremque iucundum trahit, qualem in aliis generibus aeris adfectant oleo ac sale** while the other concludes ...**oleo ac sole.** That is, the second version, which is the older, asserts that an attractive colour is produced by the action of oil and sunlight, while the first attributes it to oil and salt. One version is clearly corrupt; but which? To discover the answer by experimentation, a bronze of a composition known to be Roman was

17

treated in both ways.   In the first case, treatment with oil and salt or saltwater produced a kind of artificial patination of a green colour.   In the second case linseed oil was found to undergo polymerisation if exposed suddenly to the sun, resulting in a transparent coating of an agreeable colour.   From this we conclude that the older version is the correct one.

Finally I want to discuss the **descriptores** that are also mentioned by my colleague Dr Locher.   When Schaaber first focussed our attention on them we were at once convinced of their reality.   Of course, he had concentrated on passages where they appear very clearly; and it will be convenient if I follow him in the chapters on iron.   To begin with chapter 140, the main **descriptor** is obviously ´iron´.   The next must have been something like ´work of art´, for he lists statues and a valuable beaker.   This second **descriptor** is next replaced by a third, most probably **terra**, for he lists countries and well as soils.   Next he took **aqua** in chapters 144 and 145 , accompanied by a **descriptor** for different peoples.   Then he adds the **descriptor** **oleum** to **ferrum** and **aqua**.   An important **descriptor** for chapters 140 to 146 must have been **ignis**.   In chapters 147 and 148 the system of **descriptores** becomes particularly clear.   Here his **descriptor** is clearly **magnes lapis**, which to our understanding can be magnetite or magnesite, the first being a compound of iron, the second of magnesium.   Pliny could not distinguish between them, so he wrote: **...nescio an vitro fundendo perinde utilis, nondum enim expertus est quisquam;...**   He was right in thinking than an iron compound could never be useful in glass making.   But magnesite can be added to glass (it is magnesium carbonate).   In this way Pliny became a victim of his own otherwise effective system.   The **descriptor** for chapters 149 and 150 is **ferrugo**, after which medical things are treated, but again in the order **acies, ignis, ferrugo.**

The same system can be found in the chapters following 30 (on amber) in book 37.   The first main **descriptor** is **sucinum**; and as always at the beginning of of a new item, we find reflections on the moral and ethical significance of the item.   In 31, a new **descriptor** appears: it is **electrum**, another ancient word for amber.   In 33 there appears a new word, **liguria**, but it is not before 34 that a third Latin word for amber, **lyncurion** appears for the first time.   Pliny tries to find out what this might have to do with the animal lynx, but is not successful and reports only at second hand and without personal belief. As I have written elsewhere, the real connection is that at the northern boundary of the Appenines amber occurred naturally in antiquity, and occasionally does so today. From there it was carried to the Po (the ancient Pedanum). What Pliny did not know was that the ancient Ligurii once lived in the Po valley, before they moved to the Lingurian coast. And as the simple name **Cypricum** stands for copper, the proper name **Linguricum**, but written in Greek characters in Greek literature, once stood for the amber of that region. But to return to the **descriptores** we can see again that just as in the chapters on iron, the next **descriptor** is **terra**, used in chapters 35 to 41, and the third descriptor is obviously **sucinum**, then **electrum** and then all the other names amber has in the different countries. After reporting some fabulous stories in chapter 42, he wrote **Certum est gigni in insulis septentrionalis oceani...**, following the same pattern as we saw in the chapters on iron. Without reference to an elaborate analysis of the text, which would be inappropriate here, it is difficult to explain further how closely Pliny followed his system of **descriptores**; personally I am convinced that Pliny wrote his key-words at the top of his **tabulae ceratae**.

Chapter Three

THE STRUCTURE OF PLINY THE ELDER´S NATURAL HISTORY

A Locher

"A literary monstrosity" was what the great philologist Eduard Norden called Pliny´s **Natural History** in his history of Roman literature. And in his standard work, **Antike Kunsprosa,** still widely quoted today, he counted it "among the worst works we have, from the point of view of style".

These damning judgments determined my relationship to Pliny the Elder for a long time, and I believe I am not the only philologist who fared in this way. For many classical scholars, Pliny was simply not an author at all. German classical philology has always preferred to hold with Goethe´s Faust, who set heaven and earth in motion in order to grasp the "intellectual band" which holds everything together, than with his assiduous Adlatus Wagner, who admittedly "knows a great deal" but "would like to know everything". My own contact with this "bad" work was accordingly limited to finding in it negative examples for exercises in Latin style.

This changed suddenly in 1977 when the **Projektgruppe Plinius** (Pliny Project Group) - my colleague Rottländer describes the members and the working methods of the group in his chapter - entrusted me with the task of translating technologically significant chapters of the **Natural History.** It would be better to say: the task of preparing the way for a translation. For the insistent queries of the chemists, metallurgists, archaeologists, mineralogists and physicists who work together in this group often proved the purely philological translation, that is, the translation achieved with dictionaries and grammars, to be inadequate. For example, the technological pertinence of what is meant in the description of the process of gilding in XXXIII.65 by the words

cruciare,   exharenare, exhalare and **recoquere,** can
hardly be gathered from dictionaries. The
philologist is simply bound not to have a concrete
notion of this. However, this is the very thing
that interests technicians and technological
historians. And suddenly I understood clearly
those words which appear in the **praefatio** to the
**Natural History: Humili vulgo scripta sunt,
agricolarum, opificum turbae, denique studiorum
otiosis.**"It is written for the common people, for
the mass of peasants and artisans, and only then for
those who devote themselves to their studies at
leisure". If one takes this intention seriously,
then one can really appreciate the arrogant,
unrealistic attitude of classical philologists who
simply do not consider as literature any works which
are not exactly stylistic masterpieces. This is
how I came to see the main content and the main
value of the work in the in' the technical and
scientific information it offers. That is,
incidentally, in the majority of cases, the
perspective from which geographers, historians of
art and culture and archaeologists have always
approached the **Natural History.** This is proved by
the bibliographies.

During my translation therefore, which up to now
has always been of technologically significant parts
of the text, I gave my main attention, in accordance
with my set task, to the technological information
the text contains. Of course, I was also struck by
passages where such information was not given.

As time passed, and as the amount I read
gradually increased, I began to realise that these
passages of technological information are
interrupted by parts which repeatedly show distinct
similarities to each other, indeed, that here the
same elements continually recur in an order which
frequently varies. I believe that with this
realisation I have hit upon a peculiarity of the
structure of the **Natural History** to which too little
attention has been paid up to now.

So as to structure, the comprehensive composition
of the **Natural History** is clear and has frequently
been presented: the universe, earth, human beings,
animals, medicines from plants and animals,
minerals. This is an order, incidentally, which is
strictly non-evolutionary. Years ago (in **Arktos**
IV 1966) O Gigon wrote about the consequences in
his illuminating article "Plinius und der Zerfall
der antiken Naturwissenschaft" (Pliny and the
decline of ancient science). But now to the smaller

21

structures.   How are the individual passages built
up, how are the pieces of information arranged,   and
by   what,   if anything, are they interrupted?   This
brings us to our theme.

Only part of the text   consists   of   undocumented
factual information without further additions:

II.89: **in ipso caelo stellae repente nascuntur.**
("Stars come into being spontaneously in the sky".)

IX.89: **Polypi   binis   mensibus   conduntur.**
("Polyps remain hidden for two months".)

XIX.83: **Raphanus   fimum   odit.**   ("Carrots   (or
radishes!) hate manure")

XXXIV.149: **Ferrum accensum igni nisi duretur
ictibus, corrumpitur.**   ("If iron made red hot in
the   fire   is   not   hardened   by   striking,   it   is
spoiled".)

Whether Pliny considers this information to be so
obvious   that   it   needs   no   explanation   or
documentation, or whether he found it like that in
his   sources,   possibly a handbook, at any rate this
form of   information   predominates   in   the   medical
books,   20 - 32, where it frequently takes the form
of prescriptions.

Now to the types of information which are not of
this   form.   There is, first of all, the historical
reminiscence.   Previous events, which are   seen   in
connection   with   the   event   just   decribed,   are
narrated   from   the **acta,**   the **monumenta**   or   from
authors   of   Roman or non-Roman history mentioned by
name: in II.147 (in the context of heavenly
portents)   we   learn   of   a   series of **portenta** from
Roman history, each dated by the consuls   in   office
at   the   time: rains   of   blood,   milk and bricks.
XIII.84 gives a historical review of the history of
the **charta,**   from   reports   by Cassius Hemina up to
the   rationing   of   this   writing   material   under
Tiberius.   In   XX.199,   we learn of the suicide of
Praetor Licinius Caecina´s father with   the   aid   of
opium.   In   XXX.10 (that is, in the passage about
gold) we are given a   report   about   how   Tarquinius
Priscus   rewarded   his   son   with   a **bulla aurea** for
striking an enemy dead.   King   Diomedes,   finally,
according   to   XXXVI.21,   promised the Cnidians that
their considerable debts would   be   cancelled   -   in
return   for   the Venus of Praxiteles.   Incidentally,
Pliny is the only   source   for   a   whole   series   of
pieces of historical information like this!

In   a   considerable   number   of   passages,   Pliny
states explicitly that he was himself present as   an
eye   witness.   Thus   he   saw   with his own eyes in
II.101 St Elmo´s fire from the tips of the lances of

soldiers on guard, in IX.15 a giant whale in the harbour at Ostia, in XXXIII.63 Agrippina´s dress made of pure gold and in XIII.83 manuscripts written by the Gracchi.

Now a first retrospective survey: what I have up to now listed as continually recurring elements of the **Natural History** are the very **res et historiae et observationes** of the type that are regularly resumed at the end of the **indices,** and not only that: with the exception of books 3 - 5, they are also counted. But in contrast to the certainly stylised numbers of the **praefatio** such irregular numbers appear here that they can hardly be meant "rhetorically", as V Ferraro recently proved for the **praefatio** in the **Annali della Scuola Normale Superiore di Pisa** (1975/6). Pliny could therefore count his individual notes. Consequently, at some stage in his work, they must have existed separately. But more of that later. Let it suffice here to record that the mention of **res, historiae** and **observationes** does not offer a summary, undifferentiated survey, but to a certain extent can be found in the text as identifiable editorial categories. However, anyone who intended for example to count the **res** from out of the text would immediately fail. Pliny must therefore have had them separately in front of him beforehand. I conclude from this that the categories of the indices were not set up **a posteriori** after the text at all, but simply represent the existing **a priori** categories from Pliny´s technique of reading and making excerpts. But I shall also return to this later.

A further structural unit seems to me to be represented by the **curiosum** or **mirabile,** the amazing or grotesque incident or paradox, which, however, can in part coincide with **historiae.** Here can be found the comet of II.94 which appeared on several successive days after Caesar´s murder, always towards evening, the hunting dog presented to Alexander the Great, which made an elephant fall by circling it continuously, in VIII.150. Curious also is the love between a boy and a dolphin, in IX.25, which is carried to such lengths that every day the dolphin carries the boy from Baiae to school at Puteoli and after the early death of the boy also dies, doubtless of a broken heart.

In the discussion of pearls, we learn of what was, incidentally, probably the most expensive dessert in the history of mankind: following a bet wih Marcus Antonius, Cleopatra swallows for dessert one of the sinfully expensive pearls of her legendary ear-rings, dissolved in vinegar; that is what we read in IX.10. Or finally, the former fuller slave Clesippus Geganius, who has achieved position and dignity, reveres a bronze candelabra as if it were a cult object (XXXIV.11-12); as an additional extra to this candelabra he, humpbacked, ugly (and stinking, like all fullers) had come into the possession of the unscrupulous Gegania. She, as it says, "in shameless lust" made him carry the newly acquired candelebra naked into a banquet. To top it all, she then took him into her bed as well, and also soon into her last will and testament (what **brevitas!**). Clesippus never forgot to whom he owed his freedom - to the bronze candelabra! Incidentally it is interesting where Pliny got the story: he probably saw the tomb on the Via Appia near Tarracina between Rome and Misenum, where at least the inscribed tablet can still be seen today.

A further redactional element can be mentioned briefly. Even outside the real "medical" books, 20 -30, Pliny repeatedly points out the medical uses of different substances, most frequently in the mineral books, 33 - 37, sometimes in short marginal notes, sometimes, in several connected sentences. Here he frequently talks of the adulteration of medicaments, and just as frequently of the prices that have to be paid for them. Finally it must be added that the **auctores** who are always listed in order in the **indices** are not seldom mentioned in the text, too. This naming of auctores is dominant virtually as a structural principle in book 20 and in book 37 - the passage about amber (30-41) - just to name only the most typical passages. Significant also is the phrase **auctores habeo...** with the following accusative and infinitive, in IX.10. Pliny knows and admits that his work is a compilation.

For a moment I would like to exclude from this list of formally similar and continually recurring elements the last one, the philosophical and moral consideration, which I shall discuss at the end. So we can now survey in retrospect the fact that Pliny offers his information divided up into certain types: **res** (or **medicinae**), **historiae, observationes** as the basic division, and these are again split up into quotations from **auctores,** which one can classify as **historiae,** into **curiosa,** which can be attributed to the one or the other type, and finally into these very philosophical and moral considerations. All this comes in a varied sequence.

The structure of the Natural History

The consecutive reports of varying type are held together by subsumption under one respective key-word or combination of key-words. My colleague Rottländer also discusses this in his chapter. Here I want to discuss another aspect of the same matter. In a recently published little monograph on Pliny, the author writes in a resigned tone "we can no longer determine how Pliny proceeded" (R König/G Winkler, **Plinius der Ältere**, Darmstadt, 1979, p30). I believe that I can overcome this resignation, at least in part. The composition of the **Natural History** reveals a working method which is thoroughly thought out from a technical point of view.

Pliny the Younger gives us a very illuminating report about this method in **Epp.**3,5. He explains to us in what an optimal manner Pliny made use of his free time, time which he could ill afford to spare, as he tells us explicitly in **praefatio**, 18. He was a fanatical time-saver, short-sleeper and night-worker: even when he was in the bath or sedan chair, he used the time for reading or dictating. Reading aloud even took place during meals. All that, together with the almost grotesque details, is mentioned in this letter, which is certainly known as a whole. Paragraphs 10 and 11 however contain a detail which I consider to be revealing and which offers a hint which has been overlooked up to now about the compiling of the **Natural History**. In rapid succession we are told **liber legebatur, adnotabat excerpebatque. Nihil enim legit, quod non excerperet:** "A book was read aloud. He ´adnoted´ and made excerpts. For he never read anything without making excerpts". And in paragraph 11: **liber legebatur, adnotabatur, et quidem cursim:** "A book was read aloud, it was ´adnoted´, and that at a fast pace".

Here **adnotare** is twice distinguished from **excerpere**. **Excerpere** presents no difficulties. It means "writing down literal or summarising extracts from a text that is either read or read aloud". But why then **adnotare** as well? The German philologist A Klotz presumed as early as 1907 (**Hermes, 42**) that it could mean the marking of certain passages to be copied by Pliny´s ammenuenses. But we, I think, can go further. **Adnotare** first appears as a neologism in the sphere of philological and literary activity as late as in Seneca the Elder. It means predominantly critical or laudatory writings adding to things already written. Admittedly it is later reduced to simply meaning "noting" and at the same time takes on the even weaker meaning of "to take note of". In the archives of the Thesaurus Linguae Latinae in Munich, I examined the use of the word as found in Pliny the Younger. The result was clear: in cases where it refers to written material at all, **adnotare** always means in Pliny the Younger "making notes on something already written". **Adnotavi in urbem remittendos** it says in **Epp.**10,96,4, in the famous letter to Trajan. It refers to the official´s note that Roman citizens were to be brought before the imperial court. Pliny is obviously referring to a list submitted to him. This usage is confirmed by a large number of comparable passages.

The question remains: what notes did Pliny add to his excerpts? We are familiar with the results of his constant reading and making excerpts. In the same letter 3,5 it says in paragraph 17: **electorumque commentarios centum sexaginta mihi reliquit, opisthographos quidem et minutissimis scriptos; qua ratione multiplicetur hic numerus.** "He bequeathed to me one hundred and sixty notebooks of selections, each page written on both sides in tiny letters; this means that the number is even doubled". If Pliny the Younger´s report is correct, and we have no reasonable grounds for doubting this, then the selections in the **commentarii** were collected in the exact order that they occurred in his reading. Thus there came into being consecutively more or less compressed short extracts, for example medical ones from Sextius Niger, on the history of art and culture from Juba, mineralogical ones from Theophrastos and so on. For someone working under the pressure of time, it would have been an almost insurmountable task to filter out what was required respectively from these 160 scrolls, which would then have to be combed through thoroughly each time. We simply cannot expect such an irrational use of time of a man who was pained by the loss of "ten lines" (thus paragraph 12 of the same letter) and who scolded his nephew for going on foot because it was waste of time (the same letter, paragraph 16). It is my opinion therefore that Pliny added, or had added, to his **commentarii** notes which would make it easier to find them later. That can only have happened in two different ways, and here I must thank H Tränkle in Zurich and K Gaiser in Tübingen for a suitable hint: Pliny either dictated **adnotationes** into the **commentarii** to serve there as retrieval aids, for instance in the margin, or these **adnotationes** were added by Pliny himself in the margin of the books he read or were read to him, and contained hints as to where the excerpt in question was to be included in the **commentarii**. In this case the **commentarii** would then have been a kind of structured collection of material for the **Natural History**. Pliny would then have already been able to mention to his slaves the scrolls in which the required material was to be found in individual cases.

I believe that one can go still further. There are pointers to show what these individual **adnotationes** looked like: D Detlefsen examined in 1898 (in his **Untersuchungen zur Zusammensetzun der Naturalis Historia des Alteren Plinius** - "Examination of the Compilation of Pliny the Elder´s Natural History") the remainders of the books as well and found that Pliny had apparently worked through a considerable number of texts according to a formal set of aspects and afterwards had arranged related incidents in their own categories. It is my impression that such formal perspectives are represented in the **Natural History** which was probably worked following the same method, by the key-words and brief questions of the **indices**. **Quid sit..., Qui primus..., De..., genera, differentiae, Quando primum Romae...,** - I am waiting for a papyrus from the **commentarii** to be excavated in which these **adnotationes** stand in the margin. For they have, as I have already mentioned briefly, determined **a priori** the direction of the questions and the principle of structure.

The fact that this cannot be strictly proved is completely clear to me, but there is a definite probability in its favour. If that is so then a merit of this work, and its author, which earns our respect - in spite of all the superficiality, errors and stylistic deficiencies - becomes apparent: the merit of re-arranging a disordered, diffuse flow of information into a meaningful sequence in a very short space of time.

Here we must again point out a decisive indication of Pliny´s method of working. The countless pieces of information that Pliny worked on must have gone through a stage in the compiling when they were countable. Quite clearly there was a stage of mechanical separation. Here I imagine a pile of little tablets, of scraps of papyrus, each of which contained one passage taken from the **commentarii**, be it **res, observatio** or **historia**.

Pliny therefore created ´by hand´ an instrument which, as O Schaaber remarked years ago, worked like a modern computerised documentation system. So he could rightly praise himself: **Nemo apud nos, qui idem temptaverit, nemo apud Graecos, qui unus omnia ea tractaverit (praefatio,** 14). "There is no one among us who has attempted the same thing, no one among the Greeks who has dealt as an individual with all this together".

It is, however, in my opinion, not certain whether Pliny´s merits are really to be seen only in this sphere of technical redactional achievement. The arrangement of the individual pieces of information chosen by him perhaps also meets a higher philosophical demand: his preference for considering each and everything first of all with regard to human beings, for example the use and misuse human beings make of gold (XXXIII.1ff.) or iron (XXXIV.138ff) - is all that really so inappropriate?

One can also understand the loosely varied sequence he chooses of pieces of information, historical reminiscences, his own personal observations and, something I have excluded intentionally up to now, philosophical and moral reflections, as a pragmatic stock-taking, resulting from the need to compose, to catalogue and to make an inventory for the benefit of mankind! Admittedly this is not a philosophy, but a survey of physics and biology, anthropology, history and ethics, that is to say, a basis of discussion for a philosophy.

In these last mentioned philosophical and moral reflections, which can be found scattered throughout the whole work, sometimes in longer contexts, sometimes in short remarks, Stoic-Epicurean thinking predominates. A few examples of this are: II.117-8, where the discussion of winds gives rise to a reflection about the ethos of the sea-faring man; VII.5 contains the pessimistic consideration as man as the only being in the whole of nature to visit evil upon his own kind; in IX.105 shell fish give rise to a digression about human addiction to luxury. At XII.1-6 he reflects upon the original significance of trees for mankind; at XX.1 upon the elementary natural forces of sympathy and antipathy, while XXX.1-4 introduces the discussion of the treatment of gold with criticism of human **avaritia** as greed and desire for domination in the widest sense.

It can be read in many passages about him in the literature that Pliny is not original as a philosopher: however, this is in my opinion invalid

in one respect.   I know of no author of the
ancient world who expressed in such impressive words
the responsibility of man for nature which has been
entrusted to him, and who deplored man´s failure in
this matter.   This is perhaps where the "synthesis"
of the whole work, which so many critics miss, is to
be found.   Love of nature, which in Pliny is
undoubtedly of a religious nature, is not only
expressed in the final invocation of **parens rerum
omnium Natura,** but is already expressed in extremely
short marginal notes, two of which I would like to
mention here at the end: in XXXIII.73 Pliny has
described a method of mining gold in Spain in which
whole sides of mountains are undermined and
collapse.   He comments **spectant victores ruinam
naturae:** "In victorious pose they regard the ruin of
nature".   (It is only a question of time before the
ecology movement claims Pliny for its own.)

A second passage showing great sensibility can be
found in the introduction to the part on insects,
XII.2-4: **...in his tam parvis atque tam nullisquae
ratio, quanta vis, quam inextricabilis perfectio!:**
"In such little and insignificant creatures, what a
wealth of meaning, what force, what inextricable
perfection!"

And a little further on he continues: **sed
turrigeros elephantorum miramur umeros taurorumque
colla et truces in sublime iactus, tigrium rapinas,
leonum iubas, cum rerum natura nusquam magis quam in
minimis tota sit.**   "But we admire the shoulders of
elephants, which carry tower-like structures, the
necks of bulls and the powerful tossing of their
horns, the hunting of tigers, the manes of lions,
although nature is nowhere more present as a whole
than in the smallest creatures".

In 1951, in the **Bulletin de l´Association
Guillaume Budé,** Alfred Ernout published an article
in which he passed a moderately appreciative
judgment on Pliny and his **Natural History,** the first
person to do so, as far as I know.   At the end he
quoted at length the sombre anthroplogy of VII.5 and
regretted that mankind had not followed the
teachings set out there.   After the quotation he
wrote the words of appreciation: "Ce n´est pas si
mal dit - ni si mal pense" - "That is not put so
badly, nor thought so badly either".   This judgment
could also be passed on the **Natural History** as a
whole.   It is not thought so badly, and not
compiled so badly either.

Chapter Four

THE PERILS OF PATRIOTISM: PLINY AND ROMAN MEDICINE

V Nutton

The regular attender at the salons in the Paris of
Louis XV could not have failed to meet Mme de
Zoutelandt, the wife of the royal engineer, Seigneur
Boisson. Mme de Zoutelandt was a patriotic lady
with a bee in her bonnet, to demonstrate to the
French that the writers of her native Holland were
at least their equal in taste, talent and sound
learning. To this end she published her memoirs of
Holland and the neighbouring German courts, dabbled
in theology with **La véritable Babylone demasquée** and
translated into delicate French the celebrated
autobiography of Jan de Wit, Grand Pensionary of
Holland. Nothing daunted by the moderate reception
given to her endeavours, in 1730 she produced a
double volume of translations of Dutch medical and
religious writings from the previous century. We
are not concerned here with the theological
correspondence of Anne-Marie Schurmans, whose
precocity as well as femininity astounded her
audience, but with Mme de Zoutelandt´s other
paragon, Jan van Beverwyck, city physician of
Dordrecht. Three of his works fell victim to her
eager pen, the longest by far being his **Bergval**,
"The Avalanche" or **Défenses de la médecine contre
les calomnies de Montagne, dans le chapitre 36 de la
seconde partie des Essais.** Michel de Montaigne´s
sceptical view of the aims and achievements of the
physicians of the sixteenth century was, and is,
famous. It pointed to the doctors´ confusions in
method, in philosophy and therapeutics, and its
indictment of an apparent contract between avarice
and ignorance was so powerful that, almost fifty
years later, it could still spur a young, erudite
and talented Dutchman to spring to the defence of
medicine.

Beverwyck´s argument was cogent, his facts more

accurate than his adversary´s, and his diagnosis sound. Montaigne, he alleged, had based his attack on the validity of medicine upon tainted and untrustworthy evidence, the **Natural History** of the elder Pliny. Alas for the future of Dutch scholarship in the Parisian enlightenment, Beverwyck failed to convince the French; Mme de Zoutelandt´s project sank like a stone; and her revival of the writings of the Dutch Hippocrates escaped the notice even of his biographers.(1)

But Beverwyck was right, at least in his main contention. Montaigne did take over Pliny´s jaundiced view of medicine, but we should not be in the least surprised at this. Not only was the history of medicine, as then purveyed in such popular and universal handbooks as Ravisius Textor´s **Officina**, entirely taken from Pliny, but even the rediscovery of the Greek original texts in the middle years of the sixteenth century failed substantially to remove the Plinian bias.(2) Pliny´s prejudices linger on even in the best academic circles. The medicine of Rome still takes a very distant second place to that of Greece, as something at best derivative and at worst positively murderous. Where praise is given, it is for sewers and aqueducts, not for surgery or pharmacology, and the multifarious evidence of Galen for the medical life of Roman imperial Italy is never exploited to the full.(3) Yet this is our fault rather than Pliny´s, for it must be stated at the outset that without his evidence the historian of Roman medicine would find his work both a great deal more difficult and a great deal more tedious. It is Pliny, above all, who provides us with a framework, of both fact and interpretation, that can be used to build up a proper description and history of medicine in the late republic and early empire. Secondly, perhaps more than anywhere else in the **Natural History**, it is when Pliny deals with doctors that one can hear his authentic voice. On many topics his tone is that of a mediator of the opinions of others, yet when he discusses doctors and their therapies, his views are unmistakably his own, even where he can be shown to be taking his facts from an earlier author.

It is here too that his interest in the curious workings of humanity links closest with his own sentiments and possibly his own experiences. Who could forget his denunciation of the royal physicians, grasping and adulterous? or his sturdy opposition to trendy innovations, whether they were cold baths, the prescribing of wine or Thessalian

Methodism?(4)

The passion of his involvement is further
heightened by contrast with the calm exposition of
the development of medicine by his near-contemporary
as an encyclopedist, Cornelius Celsus, and, in
particular, by the fact that such trenchant
opposition to medicine is rare among surviving
writers of high culture. To take but two examples,
in the pages of both Cicero and Seneca, the
physician is given a relatively good press. Dr
Asclapo of Patras is described as a man of learning,
kindness and fidelity; the death of Dr Alexio is the
occasion for immense grief, proof that here was a
friend, not just a servant; and Cicero's eulogy of
Dr Glycon, "a modest and frugal man, whom no thought
of self-advantage could drive to crime", should not
disguise from us the uncomfortable fact that, when
Cicero wrote that letter, Glycon was languishing in
gaol on suspicion of murdering his patron or, at
least, of culpable negligence.(5) Closer to
Pliny's own day, the writings of Seneca continued
the Ciceronian theme. The true doctor is a man of
high morality, the purveyor of an article almost
beyond price, health and even life itself; he
endears himself to his patients, not by his
mercenary art, but by his loving kindness. He is
the intimate and helper of those who seek his aid,
even though his task is made ever more difficult by
the burgeoning of diseases of luxury.(6) As for
his own personal physician, the hypochondriac
philosopher waxes lyrical in his praise.

> He spent more than the average doctor on me; it
> was for my sake that he took precautions, not
> to preserve the reputation of his art; he sat
> beside those in distress; he was always present
> in times of crisis; no duty burdened him, none
> sickened him; he heard my groans with sympathy;
> amid a crowd of patients, my health was his
> first concern; he attended others only when my
> health permitted it; I was bound to him, not as
> to a doctor, but by ties of friendship.(7)

Pliny's view of doctors is far different
from this saccharined eloquence. Not for him the
accolade bestowed on Seneca by Karl Marx, the
celebrated professor of medicine at Göttingen in the
last century, who hailed him as a fellow
practitioner of medicine.(8) Instead, Pliny is
compared with the satirists and comedians who, from
Aristophanes through Horace, Lucillius the Greek

epigrammatist and Juvenal to Lucian and beyond, poke fun at the doctors' incompetence, greed and failure to live up to their own protestations.(9)   Such a comparison immediately diminishes the value of Pliny's arguments, for the wisecracks of the saloon bar or the gossip of the court, as titillating and as inaccurate then as now, are **a priori** unlikely to be as indicative of the truth as the personal confessions of a philosopher or of Rome's greatest orator.   Pliny is thus himself turned into a curiosity, a curmudgeonly **laudator temporis acti**, a patriotic buffoon, a crinkled Cato.

My intention in this chapter is neither to bury Pliny nor to praise him, but to try to see how far there is any truth in his contentions about the practice of medicine in his own day.   I shall not try to answer that hoary question, 'What did the Romans think of their doctors?', for, as I have already implied, one can assemble a mass of texts to support practically any position,(10) and I am always reminded of the poem of the Welsh renaissance Latinist John Owen:

> God and the doctor we alike adore,
> But only when in danger, not before.
> The danger o'er, both are alike requited
> God is forgotten and the doctor slighted.

Even with Pliny's own diatribes, one can find contradictions and qualifications which make the reconstruction of a consistent opinion far from easy.   Yet certain themes do appear in more than one book, and Pliny's overall view of contemporary physicians is clear.   He is against them.

The reasons for this opposition are many; doctors are expensive, unnecessary, incompetent; they indulge in over-subtle argumentation: they are blown along by the winds of fashion; they have brought about the deaths of the defenceless as well as the degeneration of the Roman state; they have turned Romans aside from the true, efficacious and inexpensive herbal remedies of the Roman patriarchs and of Cato and introduced them to the seductions of a medicine that was not just un-Roman.   It was Greek.

To many of these allegations I can and shall offer no defence.   Pliny is often right, even if he regularly exaggerates the points in his favour.   To take but one celebrated example, his accusations that physicians can commit homicide with complete impunity.   This claim, which in the sixteenth

century re-echoed round the world from Poland to
Mexico, if not to Peru,(11) is surprisingly true, at
least as it applied to full Roman citizens. Roman
law carefully distinguished between unforseeable
accident and incompetence, and an unskilled doctor
whose patient died as a result of his **imperitia** or
inscientia was liable to be sued for damages.(12)
There was no problem about this if the unfortunate
victim was a slave or an ex-slave, for his master
could sue the doctor under the **Lex Aquilia**, which
covered wilful damage to property - and in Roman law
a slave was as much a piece of property as a cow or
a cooking pot.(13) But what of free Roman
citizens? The traditional view, at least for
Pliny´s period, was that no one was the owner of his
own limbs and hence the **Lex Aquilia** could not and
did not apply if a free man lost a hand or the sight
of an eye.(14) No redress here, then, for the poor
unfortunate, unless the doctor happened to be a
slave or ex-slave himself and offering his services
under a contract of **locatio conductio**, a form of
hire. Then an action was possible, on the grounds
of failure to perform a contract adequately, but
here again two qualifications must be made.(15)
Many doctors seem not to have hired out their
services, but to have taken like barristers a more
professional and, one might say, a wiser stance.
They received no fees, they made no charges; their
services were given free; and any money they
received was in the form of gifts.(16) No hiring
and firing here, and accordingly no legal
recompense. Secondly, this action was only
available to the patient if he survived; his
relatives had no claim on his behalf, if he died
under care.

Things then looked black for the average Roman
citizen; but Roman legal wisdom, helpful as ever,
had another ingenious solution to hand. If an
angry cobbler knocked out the eye of a free-born
apprentice with a last for not mending shoes
properly, or if a teacher harmed a recalcitrant
free-born pupil while chastising him; or if a
neighbour´s dog or horse turned nasty and bit a
free-born son, then the head of the family, the
paterfamilias, had a claim under the **Lex Aquilia** for
damages to his son who was technically in his power,
in potestate.(17) The analogy between owning a
slave and having **potestas** over a son clearly
suggested to a resourceful lawyer a solution to one
difficulty, and arbiters and plaintiffs were glad to
concur. But what of a father, the **paterfamilias**

himself?   Alas,   Pliny   is   right.   The splendid
orotundities of the Roman Law, 'no man is the owner
of his own body'; 'no one can put a monetary value
on a free man' amount to one thing.   If the head of
the family died or suffered injury at the hands   of
the average doctor, there was nothing more to be
done; his dependents had no claim on anyone,   save
perhaps the gods.(18)
    What then of Pliny's allegations that doctors
were out to make money?(19)   Again, one can have no
hesitation about agreeing about the immense wealth
amassed by the physicians in the highest echelons of
Roman society, the doctors to the emperor and court.
Memorials   to   the   benefactions   of C Stertinius
Xenophon and his family litter the island of .Cos,
and they lend credence to Pliny's stories of their
rebuilding of parts of Naples and of Dr Crinas'
gifts to Marseilles.(20)   One can point to T
Claudius Menecrates, Nero's doctor, a   prolific
author, honoured by many famous cities,(21) or to
his contemporary, T Claudius Tyrannus, an imperial
freeman, who returned to Magnesia on the Maeander
where, laden with special privileges and "approved
by the divine judgement of the emperor", he enjoyed
the   profits   from   a   series   of **ergasteria,**
manufacturing establishments rather than surgeries,
which he set up in a nearby village.(22)   But these
are physicians to the emperors and their household,
and, in the west at least, signs of prosperity, let
alone immense wealth, among average doctors are
rare.   I   can   cite from epigraphic sources merely
one example of a medical career from rags to riches
outside court circles, that of the egregious P
Decimus P l Merula, of Assisi.   This ex-slave,
medicus clinicus chirurgus ocularius, left at his
death a total of 800,000 HS - and this after having
erected, for 30,000 HS, statues in the temple of
Hercules and, for another 37,000 HS, paved some of
the town's roads.   His fellow citizens knew a soft
touch when they saw one; he was charged 2,000 HS for
the honour of being a **servir,** and he had to pay for
his freedom 50,000 HS, over twenty times the going
rate for the period.(23)   Far more typical was
Barbius Zmaragdus, a humble member of an unimportant
collegium sacrum Martis at Aquileia, while the
presence of doctors in the registers of burial clubs
at Tibur and Latinum suggests poverty, not Pliny's
fabulous riches.(24)   Rhetorical excesses triumphed
again, and by implying that all physicians were like
the doctors to the court, who from the mists of time
onwards have always made money, Pliny tarred the

whole of a very varied social and occupational group with the same brush of avarice and adultery.(25)

What then of his allegations of widespread incompetence? Do they not combine with the melancholy laments of the tombstones to condemn the medical profession? Who could forget Euelpistus, **anima innocentissima,** whom the doctors killed; or Iulia Prisca, aged twenty, whose husband could only console himself for the mistakes of her physicians by the thought that death could carry kings away likewise; or Ephesia Rubra, a devoted mother, whose end was hastened by her physicians; or, most poignant of all, Aurelia Decia, a paragon of femine virtue, cruelly dead **per culpam curantium** at the age of twenty eight, while her husband was away from their home by the banks of the Danube.(26) What can we say of the wounded gladiators, from Spain to Bithynia, whom their physicians failed to cure?(27) or of the public slave Felix, whose sight was restored after ten months by the intervention of the **Bona Dea,** when he had been given up by the doctors?(28) Do not these miserable memorials prove the truth of Pliny's complaint? Again I plead for a reserved verdict. Such calamities might occur at any time, no matter how competent the doctor or dutiful the nursing; for death, one might say, was the occupational hazard of being a patient. Besides, the failure of the doctor is always more striking and fraught with greater social consequences than, say, the incompetence of a cobbler or carpenter. The collapse of a chair is usually less disastrous than that of a patient. Secondly, in Rome, that teeming metropolis, the normal social constraints that governed a doctor's conduct and behaviour in the face-to-face society of small provincial towns were far less pressing and the supervision of physicians was far more remote than in the minature democracies of Asia Minor.(29) From that perspective, the move of a doctor to Rome could be seen as a flight to escape punishment for failure, a resort to the anonymity of the big city, indeed, almost as a confession of crime - this, I hasten to add, an opinion put forward by Galen, himself an immigrant physician, although his move to Rome, he claimed, was involuntary. He was forced to leave Pergamum to avoid the envy and malice of his fellow practitioners there.(30)

Not all the doctors themselves were unaware of the consequences of failure. From Hippocratic times onwards, writers on medicine ordered the physician to treat only those conditions he believed

he could cure, and to leave the rest well alone.
"To help, not to harm", a Hippocratic aphorism,
carried with it the implication that the doctor need
not treat all of those who called on him. The good
doctor was not moved by the expectation of the
unlearned; it was safer, as well as better for his
own reputation, for the doctor to refuse to
intervene rather than attempt a cure beyond his, or
even anyone else's, capabilities.(31) If
amelioration was possible, then he might act; but,
at least in his own eyes, no blame was attached to
inaction. Stobaeus reports, with apparent
approval, the **bon mot** of Herophilus, who defined the
ideal doctor as the man who could tell
(**diaginoskein**) what could be done, and what not -
and by implication when to leave well alone.(32)
Such an ethic, as well as considerations of
reputation and self-preservation, may well have
reduced the temptation to intervene at all costs, to
rescue the patient, no matter what, and to brandish
the knife more in desperation than in hope.

But it is with Pliny's contention that medicine
in Rome was both unnecessary and un-Roman that I
wish to take the greatest issue, for it is here that
Pliny's influence has been felt most strongly in
subsequent discussions, and where he himself reveals
most self-contradiction.(33) On his main point
however, that medicine in his day was practised
largely by Greeks, all sources, both literary and
epigraphic, concur. A rough estimate, based on the
dating of Hermann Gummerus, suggests that among
doctors recorded on inscriptions in the West, in the
first century AD, only 17 out of 176 show no signs
of Greekness, that is, less than ten percent; the
figure rises in the second centry to 51 out of 202,
about 25 percent, and possibly to 35 percent in the
third century, 21 out of 61. If one looks at
particular areas, my sample strongly bears out
Pliny's arguments for Rome, where over these three
centuries some 93 percent of named doctors bear
Greek names - for Italy outside Rome, for Provence
and Spain, the percentage drops to around 70.(34)
For my purposes it makes no difference whether these
names were borne by Greeks or by non-Greeks
masquerading as Greeks. The pattern of expectation
was the same. With the imperial physicians, we
find a similar story; most are Greeks from Greece,
Asia Minor or that little Hellas beyond the Alps,
Marseilles.(35) Their compatriots flocked to Rome
from all over the Greek world, from Lycia and
Pamphylia, from Smyrna, Tralles, Ephesus, Nicomedia,

Nicaea and Thyatira; and a new inscription reveals an immigrant oculist from Thebes.(36)   They would appear to have been welcomed; and they obviously thought that they had a saleable commodity.   But were they fooled, or fooling?   Were they indeed the harbingers of death and the transports of degeneration?   Was the introduction of Greek medicine into Rome a cause or a consequence of Hellenisation?

Pliny's eloquence is here powerful, especially when combined with the biting prose of the elder Cato.   But, as professor Astin has taught us,(37) Cato's anti-Hellenism, his McCarthyite opposition to all un-Roman activities, owes more to his appreciation of political advantage than to any devotion to the truth, and we should at least have our suspicions about such a witness.   Nor is the evidence of Cassius Hemina all that Pliny claims, for there is an easy explanation to hand to account for the career of Archagathus the Peloponnesian.(38) It was he who in 219 BC became the first ever physician in Rome, according to Pliny, and his subsequent failures earned him the soubriquet of "Executioner" and a reputation in later centuries for utter incompetence, even among writers who could not spell his name correctly.(39)   He returned to Greece to public derision and, we might assume from Pliny's comments, set back the progress of Greek medicine in Rome for a generation or more.

This oft-told tale needs serious correction. Rome herself was becoming open to Greek influences increasingly from 320 BC onwards: and the introduction of the cult of Asclepius to Rome in 292 BC is only one example of a general trend.(40) Scattered pieces of evidence suggest the presence of doctors in Rome before the fateful year of 219, even if we need not follow Dionysius of Halicarnassus in assuming that their numbers were large and their accessibility great.(41) Archagathus indeed marks a further step along Rome's road to Hellenisation, but in one special way.   The fact that he was given citizen rights as well as a surgery at public expense suggested to Louis Cohn-Haft the correct explanation: the Roman senate was acting like the council of a Greek city in sending for and paying a resident civic physician.(42)   This was the first time the state had taken upon itself to employ a health official, an index of how far the leading citizens saw themselves as already in a Greek tradition.   What, too, of Archagathus?   Was he really an incompetent butcher, driven out by popular complaint and public obloquy? Quite possibly, for Rome was not in 219 BC the magnet for the aspiring Greek that it was to become two generations later. It was very much in the backwoods, and its pull might not have been so strong as to attract the best practitioners. The motives of the frontier doctor may not always coincide with those of the Harley Street physician, and his treatment is not necessarily the most efficacious. Yet Archagathus´ return to the Peloponnese may have also been occasioned by the ending of his short-term contract or by a desire to see again the bright lights of Athens or Corinth.

I would also set the growing usurpation of Roman herbal medicine by Greek theories and therapies in the same context of a growing urbanisation of Rome itself. Traditional ties were being broken; the village herbalist could no longer service a metropolis; the drugs to be got from suburban fields could not in themselves cure the whole population of Rome. Pliny was perhaps aware of this when he enjoined his audience to get up and tramp over the fields and moors in search of herbs, but, significantly, his hero, the centenarian Antonius Castor, pottering daily round his garden, shows the limitations of this appeal.(43) Such indulgence was possible only for the wealthy who could afford their gardens; Pliny´s advice was far less applicable to the myriads in the Roman **insulae**, who lacked the basic conditions for health, let alone a flowerbed. They had to rely on the services and experience of others, on the diagnostic skills of the physicians and the entrepreneurial abilities of the drug sellers. These purveyors of health filled a need, and the growth and expansion of Roman power, particularly in the Greek East and beyond, enabled them to introduce to their clients an ever greater range of drugs, herbs and spices. The increasing wealth of the Roman upper classes meant that they could buy exotica, and they came increasingly to rely on imports or, at a lower level, on local imitations of the real thing.(44) Particularly in the early Empire there seems to have been a new opening-up of the drug trade with India;(45) and it is not just Pliny, but even Galen, who complains that the abundance of drugs from all over the world that came to the Roman metropolis caused the drug dealers to neglect the plants and herbs that grew at the very gates of the city.(46)

The rise of population in the city itself would also have contributed to a phenomenon occasionally mentioned by medical and non-medical authors, the growth of new diseases. Pliny, at the opening of book XXVI, noted the arrival of skin diseases like **mentagra, elephantiasis** and **lues**, whatever their modern equivalents are, and of an internal disease called **colum**,(47) while Plutarch could maintain that the Roman world had now moved from diseases of want and starvation to diseases of satiety and luxury.(48) This might be true of the Roman upper classes, but for many poor in the slums, it was only the corn doles that gave them some sort of protection against malnutrition and famine.(49)

In such a situation, any help was perhaps better than none, and however much one might deplore the charlatanry of certain physicians or the adulteration of drugs by certain herbalists, themes for complaint among all writers of all areas,(50) they at least held out the possibility of amelioration, and we should not be surprised to find that it was eagerly grasped by the sick.

My argument so far is that the introduction of Greek physicians into Rome was a consequence of the transformation of that city into the "epitome of the whole world".(51) Physicians were not, as Pliny argued, themselves the cause of the decline, but rather a concomitant of a profound social and intellectual change. Yet this appeal to historical inevitability cannot by itself refute Pliny´s other point, that the medicine and the therapies that the physicians brought with them were worse than useless and expensive to boot.(52) Indeed, any definite answer to this charge is impossible, for we lack detailed knowledge of so many of the essential elements for a satisfactory defence. My reply thus can only be schematic, especially since I have neither the time nor the expertise to relate in detail the pharmacological properties of Pliny´s recommended drugs.

It must be admitted at once that the period for which Pliny is our prime source, roughly from 250 BC to 70 AD, is a particularly dismal one for medicine. Leaving aside Celsus and some late deontological writings in the Hippocratic corpus, only four medical authors survive with even one of their works intact, and it would be a foolish man who built up a system based on the very diverse offerings of Apollonius of Citium, Erotian, Dioscorides and Scribonius Largus. The great days of Alexandrian medicine were over, although Markwart Michler has drawn attention to the continued prowess and reputation of Alexandrian surgery.(53) Anatomical research on the model of Herophilus and Erasistratus had fallen by the wayside, and even at Alexandria anatomy had become little more than a mere recital of bones of the skeleton.(54) Hippocratic medical theory, the belief in the four elements and the four humours, still flourished and was gradually being extended in Pliny´s lifetime by such worthies as Rufus of Ephesus,(55) but it faced stiff competition from an atomist theory of man, first put forward by Asclepiades of Bithynia in the early first century BC and later extended and and expanded by Themison of Laodicea and particularly Thessalus of Tralles.(56) This Methodism, as it became known, and particularly its champion, Thessalus, received a bad press from both Pliny and Galen.(57) Its refusal to investigate causes, its claim to be treating only a limited range of common conditions caused by a disturbance of the body´s constituent atoms and pores, which could be easily recognised from the patient´s symptoms and easily taught to others, aroused the anger of learned diagnosticians.(58) They hurled forth a barrage of abuse, more suited to the hustings than the bedside, and, in Thessalus, Pliny´s own contemporary, they created a medical bogeyman with which to terrify patients and pupils alike.(59) Of the quality of Methodist therapeutics little is recorded, but there is some indication that they were at times superior to those of their opponents. The treatment of chronic and mental disorders, if Drabkin is right, was more sophisticated and probably more effective than that offered either by the Hippocratic or by the Empiric physicians of the time.(60) Even Pliny could find a moment´s praise for Dr Asclepiades, at least for his ideals of acting swiftly, safely and pleasantly - although he had his doubts about the latter as being the introduction of yet more self-indulgent luxury into the medical world of Rome.(61)

As a counterpart to Pliny one can also set his predecessor as an encyclopedist, the unjustly neglected Cornelius Celsus, whose account of the medical controversies of the preceding centuries presents a markedly different picture.(62) His emphasis on the calm efficiency of medicine as a whole, whether what is required is treatment by diet, drugs or the most difficult and dangerous of all, surgery.(63) He is aware of drawbacks as well as advantages, and his tone and manner are moderate throughout. Yet he is not a mere translator or transmitter of Greek medical sources - he is prepared at times to make his personal preferences known and to record his own observations. He favours the treatment of the individual rather than simple reliance on a few common conditions, which he dismisses as the medicine of the slave hospital,(64) and his description of surgical operations has deservedly drawn praise from later writers on the subject. His operation for the removal of a bladder stone was still practised in the last century, and was far more effective and less time consuming than some of its medieval successors.(65) Yet, even when we acknowledge the merits of his proposed treatments, it is impossible to know how often they were attempted, and how often they succeeded. Celsus´ lithotomy, although perfectly acceptable in itself, must have led to many deaths simply through the infection of the wound, although as Guido Majno has argued, the Romans knew the antiseptic properties of wine and vinegar, and the belief that all operations were inevitably failures is a considerable exaggeration.(66)

What, though, of conditions that did not require surgery, the dysenteries, colds, and fevers that beset the average Roman? Here Celsus, Dioscorides and Scribonius Largus agree with Pliny in the range and type of their therapies. But their confidence in the efficacy of their own remedies is no substitute for proof, and this is very hard to come by, for many reasons. First, most diseases are self-limiting, and a patient would in many cases have recovered without any medication whatsoever. The drugs and diet prescribed by the ancient doctor might have speeded up the healing process, or have had little effect, or even, in a few cases, hindered a recovery, yet these distinctions could not easily have been made accurately in antiquity, and it is impossible to say whether any improvement in a patient´s condition was due to the pharmacological properties of a particular drug or to the workings of the placebo effect.(67) Furthermore, once a remedy had been followed by recovery, for whatever reasons, it would be recorded in the drug handbooks, where its success would be sanctified by age, and any apparent failure to the patient´s irresponsibility, temperament or circumstances. To challenge the supposed efficacy of a drug was to cast in doubt the results of generations of successful practice, no light task when so many alternative sources of error presented themselves.

Secondly, even if a doctor was right to asssume that a particular remedy actually helped the patient, the growing fondness for polypharmacy meant that a precise identification of the active ingredient or ingredients in a particular drug was almost impossible. To decide which of the sixty four ingredients of the famous theriac of Andromachus, Pliny´s contemporary, actually worked for the benefit of the patient was beyond the competence of a Galen, and pharmacologists acknowledged this almost openly when they talked of drugs that worked through the property of their total substance, which could not be gauged from a knowledge of the properties of their elemental constituents.(68)

There were and are further pitfalls in estimating the effectiveness of ancient therapies. Even if the chemical properties of ancient plants and herbs can be established by modern methods, the identification of a disease in a specific patient was not easy for an ancient physician. There was a confusion of terminology for both diseases and parts of the body, and categories which might have seemed clear in antiquity, when passed to us through only a small body of texts, become almost useless for the purposes of identification. Some descriptions seem to us impossibly vague: the classification of fevers by their periodicity is at best crude and at its worst misleading.(69) Conversely, in skin diseases the terminology is frequently precise but cannot easily be reconciled with the findings of modern dermatology.(70) The modern observer also has to believe that the diagnosis of the ancient doctor was correct and his report accurate, and the drugs supplied by the pharmacist unadulterated and correctly measured, which, to judge from the adverse comments of Galen as well as Pliny, may represent an unjustified leap of faith.(71)

A direct rebuttal, then, of Pliny´s claims that the therapies of contemporary physicians did not work is fraught with difficulties, yet two considerations suggest that once again rhetorical exaggeration has triumphed over truth. The first is that the problems about which Pliny complains were recognised and discussed among the doctors themselves. Two texts from the Hippocratic corpus, **Decorum** and **Precepts** which are of Hellenistic date, show authors who are trying to formulate an ethical programme for the medical profession,(72) while Scribonius Largus, who came to Britain with the armies of the emperor Claudius, devotes part of his preface to his drug handbook, the **Conpositiones**, to an exposition of medical ethics.(73) At worst, then, some physicians were aware of the inadequacies of their fellows, and we may assume that they themselves wished to do better.

Second, all ancient medical writers emphasise the importance of good nursing, attending to the patient´s needs but keeping a balance between servility and despotism. Galen places at least as much emphasis on securing the co-operation of the patient and his attendants in the battle against disease as on the actual drug therapy.(74) For many conditions this would by far be the best treatment, especially if the doctor succeeded in keeping away unwanted or officious callers. The Methodists, too, paid great attention to convalescence as an integral part of treatment, and some of the diets and activities recommended by Ascelpiades would not be out of place in a modern hospital.(75)

The activities of the Greek physicians may, then, have been less disastrous than Pliny alleges, even if less effective than they themselves claimed. This feebly favourable conclusion, vague and impressionistic as it is, may perhaps be strengthened considerably if we examine the alternatives that Pliny himself has to offer. If these turn out to be even less effective or more dangerous, then the therapies of the **medici**, despite all Pliny´s strictures, may have been the best choice available.

Pliny´s contribution to the medical literature of antiquity is twofold, the advocacy of a return to a traditional Roman rural self-help, modified to accommodate a Roman gentleman like Castor, and secondly the massive importation into scientific literature of a whole range of medico-magical remedies associated with such names as Ostanes and Pamphilus.(76) Of the latter I shall say nothing, for Pliny himself is as scathing of magic, at least in theory, as he is of Greek medicine, and contemporaries like Scribonius Largus were quite capable of drawing a line between proper medicine and superstitious nonsense, and even of rejecting sympathetic remedies that might appear to work, on the grounds that they "fell outside the profession of a doctor".(77)

But what of Roman agrarian medicine as an alternative to Greek sophistry? Even this had its drawbacks. On Pliny´s own account, this true medicine had begun to be corrupted in Hesiodic times, (78) and when our Roman sources begin to reveal the medicine of provincial Italy, the tale they tell does not inspire confidence. The average

village healer was viewed with trepidation rather
than affection. The most famous of all the native
practitioners, the Marsi, the men of the Abruzzi
mountains, untouched by Greek influence, were a race
apart. They might tour Italy with their snakes and
their simples, and advise on the making of theriac,
but they would not be welcome guests at a dinner
party. Their foreign equivalents, the Psylli and
the Nasamones, had an equally dubious reputation and
woe betide the unwitting visitor to the Paphlagonian
backwoods, for instead of receiving treatment for
his illness he might be struck dead by the glance of
an angry Palaeotheban.(79) Or take the most famous
of the medical wanderers of central Italy, Lucius
Clodius of Ancona, known to us from Cicero's **Pro
Cluentio.** He was no paragon of piety, no homespun
Harvey, but a man constantly on the move, a
circulator, hurrying from one market and fair to
another, cursorily treating the peasantry for their
ills. When he visited an elderly lady sick in bed
at Larinum, his tonic was designed to kill, not to
cure: one dose carried her off, and Clodius hurried
on his way, 2,000 HS the richer.(80)

The medical life of the countryside was no idyll.
Nature did not provide remedies of every sort to
hand for every ill. Charms and cabbage might be employed for a variety of diseases but their efficacy
was at least as uncertain as the prescriptions of an
urban doctor. This is why Dio recommended peasants
to take advantage of any passing physician and
others emphasised the importance to the traveller of
having a doctor as a companion,or, failing that,
taking along a good practical handbook of
medicine.(81) Nor would rural remedies in any way
benefit the townsman in Rome, far from fields and
friendly rootcutters. At best Pliny's strictures
might encourage the inhabitants of small towns to
seek an alternative to the local druggists, whose
reputation was not always healthy. At Capua, their
trading quarter, the Seplasia, was notorious
throughout Italy, and if we may believe Valerius
Maximus, its wares brought about the downfall of
Carthage.(82) Hannibal's victorius troops after
Cannae descended with glee on the perfumes and
spices of the Seplasia and thereafter never won a
battle. The paint was mightier than the sword.
Yet how many Romans, one might ask, actually
followed Pliny's advice? Very few, I imagine, our
author not among them, for, despite his
protestations, there is every sign that his
knowledge of medicine comes almost entirely from

books,(83) and his acquaintance with doctors, at least in part, from the Latin equivalents of the gossip columns.

My conclusions, therefore, are sombre. If one followed Pliny's advice and avoided doctors, one might indeed remain healthy, possibly wiser and certainly richer. But, if one had the misfortune to fall ill, and to be medicated from some of his fantastic pharmacopoeia, one's chances of survival were, I suspect, less than if one had been attended by Galen, or a landowner like Celsus, or had been given adequate nursing, including for good cheer an occasional draught of wine, as prescribed by Asclepiades.(84)

Almost sixty years ago, J Wight Duff delivered himself of a magnificent **sententia**, that, with one emendation, can serve to sum up this paper. "Nowehere" thundered Duff, in tones that evoke a vanished age of English imperial education, "Nowhere is Pliny more exasperating than in his maltreatment of the ablative case".(85) For the unfortunate ablative, read "The medical profession".

NOTES

1.   J van Beverwyck, **Eloge de la medecine et de la chirurgie.   Defense de la medecine...**, Paris, V Rebuffe, 1730;   A M Schurmans, **Lettres...**, Paris, V Rebuffe, 1730, which contains, unpaginated, a brief biography of the translator. There is no reference to this edition, a copy of which exists in the Wellcome Library, in the standard works on Beverwyck;   J   Banga, **Geschiedenis van de geneeskunde...in Nederland,** 1868, repr., Schiedam, Interbook, 1975, pp.286-313;   E D Baumann, **Johan van Beverwijck in leven en werken geschetst,** Dordrecht, J P Revers, 1910; or, more surprisingly in A J J Van de Velde, "Bio-bibliographische aanteekeningen over Johan van Beverwyck (1594-1647)", **Kon. Vlaamsche Academie voor Taal- en Letterkunde Verslagen en Mededeelingen,** 1932, pp. 71-121; 1933, pp.36-47, or in G G Ellerbroek, "Un adversaire hollandaise de Montaigne: Johan van Beverwijck", **Neophilologus 31,** 1947, pp.2-8, who also fails to note the complexities and changes in the various versions and editions of this work.   It is mentioned in passing by M Dréano, **La renommée de Montaigne en France au XVIIIe siècle,** Angers, Editions de l'Ouest, 1952, p.94.

2.   Ravisius Textor, **Officina,** Paris, R Chauldiere, 1520, and often reprinted: I quote its

list of **medici** from the edition of J A Julianus, Venice, 1617, pp. 332-335: cf. also his lists of magi, and of poisoners, pp. 285-7. The list in T Zwinger, **Theatrum Humanae Vitae**, Basle, E Episcopius, 1586-7, pp. 1232-1238, adds more Galen but retains the Plinian base. The next century saw two attempts to rebut Pliny; J Filésac, **Medicina Defensa**, Paris, 1618; and Georg Kirsten, **De Medicinae Dignitate contra Plinium et Platonem**, Stettin, 1647.

3.   Pliny´s account is followed closely and his prejudices, if anything, extended by the authors of the only two monographs on Roman medicine in English, T C Allbutt, **Greek Medicine in Rome**, London, Macmillan, 1921; and J Scarborough, **Roman Medicine**, London, Thames and Hudson, 1969. More recent studies, which have sought to go beyond Pliny, include: G Baader, "Der ärztliche Stand in der romischen Republik", **Acta Conventus XI ´Eirene´**, 1968, Warsaw, 1971, pp. 7-17; "Der ärztliche Stand in der Antike", **Jahrbuch der Universitat Dusseldorf**, 1977/8, pp. 301-315; K D Fisher, "Zur Entwicklung des ärtzlichen Standes im romischen Kaiserreich", **Medizinhistoriches Journal 14**, 1979, pp. 165-175; J H Phillips, "The emergence of the Greek medical profession in the Roman republic", **Trans. Stud. Coll. Phys. Philadelphia**, ser. 5.2, 1980, pp.267-275. An alternative approach is outlined by me in a forthcoming article, "Verso una storia soziale della medicina antica".

4.   NH XXIX.5.10; XIV.28.143; XXIX.5.9.

5.   Asclapo, **Ad Fam.** XIII.20; XVI.4; XVI.9: Alexio, **Ad Atticum**, XV.1-3; Glycon, doctor to Pansa, **Ad Brutum** I.6.

6.   Seneca, **De Benef. III.35.4; VI.15-16; Ep.**95, cf **Ep.**14.15.

7.   **De Benef.** VI.15.4. Is this the Statius Annaeus of Tacitis, **Annals** XV.64? On the various illnesses of Seneca, modern medical men have made play; R Neveu, "La médecine et les médecines dans l´oeuvre de Sénèque", **Bull.Soc.fr.hist.med.7** 1908, pp.163-174; M Mattioli, "La malattia acuta descritta da Lucio Annaeo Seneca: infarto del miocardio o angina pectoris?", **Cardiol.Prat.** 24, 1973, pp.147-154; M Rozelaar, "Seneca - a new approach to his personality", **Psychiatry** 36, 1973, pp.82-93. The results of their diagnoses are not impressive.

8.   K H F Marx, in a review in **Göttinger Gelehrte Anzeigen** 1877, p.16. Others have followed Marx´s lead: R Neveu, "Les théories médicales et médico-sociale des Sénèque* et de Van der Marsch",

Atti della Riunione sociale della Soc. it. stor. med., 1942, pp. 351-5; K de Caprariis, "Considerazione su alcuni aspetti precorritori del pensiero di Seneca in campo psichiatrico", Acta med. hist. patav., 14, 1967-8, pp. 73-83; K Dieckhöfer, "Aspekt einer rudimentaren Psychologie, Psycho-pathologie und Psychotherapie bei Seneca", Actes 25 Int. Congr. Hist. Med., 1976, pp 579-595.

9. Aristophanes, Ach. 1032, cf L Gil, I R Alfageme, "La figura del medico en la comedia ática", Cuadernos de fil. clas. 3 1972, pp. 35-91; Horace, Sat. II.3; Lucillius, Anth. Pal. XI.89, 131, cf P Ehrhardt, Satirische Epigramme auf Arzte, Diss., Erlangen, 1974; Juvenal, Sat. V, cf. E F Cordell, "Medicine and doctors in Juvenal", Bull. Johns Hopkins Hosp., 14, 1903 pp. 283-287; Lucian, Quomodo hist. 16.24, cf. J D Rolleston, "Lucian and medicine", Proc. Roy. Soc. Med., 8, 1915, pp. 49-58, 72-84; Anon., Timarion, cf O Temkin, The Double Face of Janus, Baltimore, Johns Hopkins U.P., 1977, p. 221f. A useful recent attempt to penetrate the fog of rhetoric is I Mazzini, "Le accuse contro i medici nella letteratura latini ed il loro fondamento", Quaderni linguistici e filologici, 1982-4, pp. 75-90.

10. H H Huxley, "Greek doctor and Roman patient", Greece and Rome, 26, 1957, pp. 132-138; A Gervais, "Que pensait-on des médecins dans l'ancienne Rome?", Bull. Assoc. Guillaume Budé, n.s. 4, 1964, pp.197-231, and the more recent works cited in n.3 above, mark little improvement on older works such as L Friedlander, De medicorum apud Romanos condicione, Diss., Königsberg, 1865, and even J Spon, Miscellanea eruditae antiquitatis, Lyons, Huguetan frères, 1685. Spon's comments a generation later brought forth in England a debate, with admittedly more heat than light, which exemplifies the difficulties of answering such a question on the basis of a few authors; R Meade, Oratio anniversaria Harveiana habita 1723; adjecta est Dissertatio de Nummis quibusdam in Smyrnaeis in Medicorum honorem percussis, London, E Buckley, 1724; C Middleton, De Medicorum apud Veteres Romanos degentium conditione quae contra J. Spon et R. Meadium servilem atque ignobilem fuisse ostenditur, Cambridge, E Jeffery, 1726; J Letherland, In Dissertationem... de Medicorum... Conditione... Animadversio brevis, London, J Noon, R Ford, 1727; C Middleton, Dissertationis de Medicorum... Conditione... Defensio, pars prima, Cambridge, E Jeffery, 1727. In all this, the

evidence of Pliny plays a major part. In **The Medical Profession in the Roman Empire, from Augustus to Justinian,** Diss., Cambridge, 1970, esp.pp. 67-86, I attempted to see how far one could use the epigraphic evidence to provide a check on the bias of the literary sources, but such an approach is impossible for the Republican period, for which only a handful of inscriptions survive; **CIL** I 707, **ILS** 7791, **Epigraphica** 34 1972, pp. 105-130 and possibly the two bilingual inscriptions, **CIL** XI 1979, 1980.˙ It should also be stressed that, before the ˌage of Cicero, we have almost no surviving sources that tell us about the general social conditions in Rome, and there is thus a tendency to rely even more on Pliny´s comments on doctors and, with far less reason, to assume that the silence of those sources that survive is of major significance. An **argumento ex silentio** is here very weak. Cf. V Nutton, "Murders and miracles: lay attitudes towards medicine in Classical Antiquity", in R S Porter, ed., **Patients and Practitioners: Lay Perceptions of Medicine in pre-industrial Society,** Cambridge University Press, 1985.

    11.   Pliny, **NH** XXIX.8.18. In the sixteenth century this phrase and many others from the same section were often used in polemics by learned physicians against those whom they considered quacks, for example, Johannes Lange, a Silesian physician, later doctor at Heidelberg, **Epistolae Medicinales,** Hanover, Wechel, 1605, I. pf.; II.47; James VI of Scotland, D Hamilton, **The Healers,** Edinburgh, Canongate, 1981, pp. 64-67; the council of Mexico City, J T Lanning, **Pedro della Torre,** Baton Rouge, Louisiana State University Press, 1974, p. 7. Cf. also F Kudlien, "Medical ethics and popular ethics", **Clio Medica** 5, 1970, pp. 91-121, esp. pp. 97-107.

    12.   **Dig.** 1.18.6.7; 9.2.7.8; 9.2.8.pr.; 9.2.52.pr.; **Inst.** 4.3.6-7. The question of skill was also to be regularly reviewed by the town council in considering whether to re-appoint or re-employ a civic doctor. **Dig.**50.9.1.

    13.   K H Below, **Der Arzt im römischen Recht,** Munich, C H Beck, 1953, pp. 109-118

    14.   **Dig.** 9.2.13.pr., with note 18 below. If a free citizen were injured as a result of deliberate fraud or malpractice on the doctor´s part, then the criminal law could be put into effect and a man gain pecuniary restitution for the doctor´s crimes, cf. **Dig.** 50.13.3. Certain

'medical' activities, such as the performance of castration, and the supplying of "deadly drugs", including love potions and abortifacients, might also be punished as criminal, cf Below, op. cit., pp. 122-134.

15.   Below, op. cit., pp. 83-98, 108f.

16.   Like Galen, who although he claims never to have charged fees, on at least one occasion received a substantial sum as a gift, XIV 647 K. Below, op. cit., pp. 81-107, argues that payment was made through a **locatio conductio**, not through **mandatum**, but both the relevent texts, **Dig.** 9.2.7.8 and 50.13.1, have been suspected as late interpolations, and may be incompatible. The view of Alan Watson, **The Contract of Mandate in Roman Law**, Oxford, Clarendon Press, 1961, p. 99f., that neither of these forms of action were possible, and that there was a sort of **tertium quid**, whose details are unknown to us, seems unlikely in view of the number of arguments on this point that are presented in the Digest.   I prefer to follow D Daube, in a review of Below, **Journ. Rom. Stud.**, 45, 1955, p.179f., who argues that although mandate is theoretically inappropriate, as being the **gratuitous** performance of an act on behalf of another, the doctor behaves as if it is, for reasons, perhaps, of status and social esteem.   Cf also on this vexed problem J Macqueron, **Le travail des hommes libres**, Diss., Aix, 1964, pp. 158-184; J Visky, "La qualifica della medicina e dell'architettura nelle fonti del diritto romano", **Iura** 10, 1959, pp. 24-66; J A Crook, **Law and Life of Rome**, London, Thames and Hudson, 1967, p. 204f.

17.   **Dig.** 9.2.5.3 and 9.2.7.pr.; 9.1.3.

18.   The only text to suggest anything to the contrary, **Dig.** 9.2.13.pr., ascribed to Ulpian (fl.210 AD) offers an action to a free man only by analogy with the **Lex Aquilia**, and has been universally suspected as a post-classical interpolation of late antiquity: see Crook, op.cit., pp. 199, 321 (for bibliography).

19.   Principally in **NH** XXIX; e.g. 2.4; 5.7f; 7.14 (Cato); 8.16

20.   **NH** XXIX.5.9 (Crinas); 5.7f.(Xenophon); although the figures for the latter's riches are uncertain, they were immense, and as H G Pflaum remarked,   **Les carrières procuratoriennes équestres sous le haut-empire romain,** Paris, Institut franc. d'archéologie de Beyrouth, 1960, p. 43, far exceeded the notional salaries of other comparable officials.

For discussions of Xenophon and his family, see
R Herzog, "Nikias und Xenophon von Kos", **Hist.
Zeitschrift** 125, 1922, pp. 189-247; A Maiuri, **Nuova
Silloge epigraphica di Rodi e Cos**, Florence, 1925,
nos. 476-478, bringing the total of Coan
inscriptions then published up to 14, and p.
176; Pflaum, op. cit., pp. 42-44. His house in
Rome was on the prestigious Caelian, **CIL** XV2.7544,
cf. **ILS** 1841. The suggestion of G
Pugliese-Carratelli that he was involved in the
refurbishment of a medical institution at Elea in S.
Italy, "Culti e dottrine religiose in Magna
Graecia", **La parola del passato** 18, 1965, p. 27,
although attractive, cannot be substantiated, and
the scale of the work carried out would be more
appropriate to a local benefactor rather than a
cosmopolitan plutocrat.

21.   **IG** XIV 1759.   He is probably not to be
identified with Menecrates of Sosandra; see P
Herrman, **Tituli Asiae Minoris**, V, 1981, n. 650, with
a full bibliography.

22.   **Inschriften von Magnesien** 113.   For
ergasterion as a surgery, see **SEG** 1175 (3rd century
BC); Galen, X.682 K.; John Chrysostom, **Patr. graec.**
58.779; 60.365, but it is more likely to mean a
manufacturing establishment, cf., from neighbouring
areas, **Inschr. Miletus**, 1, n.225; **Abh. Akad. Wiss.**
Wien, phil.-hist. Kl., 1910, p. 33, n. 52.

23.   **ILS** 5369, 7812.   Cf. R P Duncan-Jones,
"An epigraphic survey of costs in Roman Italy",
Proc. Brit. School Rome 33 1965, p.293.

24.   P Carraci, "Medici e medicina in Aquileia
romana", Aquileia nostra 35, 1964, p. 93; **CIL**
XIV.3350, 3641; **CIL** IX 740, but the restoration is
doubtful.

25.   It is worth noting that the first truly
historical record of a doctor's career in Greece,
that of Democedes of Croton (Herodotus, **Hist.**
III.131ff.) is also of a court doctor who became
wealthy, but, for various reasons, Herodotus'
interpretation differs **toto caelo** from that of
Pliny. The source for Pliny's denunciations of
Eudemus and Valens, **NH** XXIX.5.8; 8.20f., is much
more likely to have been gossip than a historical
writer.

26.   **Notizie degli Scavi**, 1911, p. 170:
L'Annee Epigr., 1952, n. 16; **Carm. Lat. Epigr.**, 94;

27. **CIL** III 4314, 12925, 14188.   Cf Galen,
XIII.600 K., an exaggeration, as shown by his own

admission, cited in **Sudhoffs Archiv** 22, 1929, p. 77.

28. CIL VI.68. I am not convinced by the attempt of A M Verilhac, "Une victime des médecines?", **Mem. Centre Jean Palerne** 3, 1982, pp. 159-161, to discover a victim of Asclepiadean wine therapy in an inscription from Chalcis. Cf. the recently published inscription from the Lebanon recording the case of a man who had seen thirty six doctors in vain, and who was cured by a god: see P Roesch, "Médecins publics dans les cités grecques", **Hist. Sciences médicales** 18, 1984, p.290.

29. On the differences between the provincial cities and Rome in their granting of tax immunity, see my articles, "Archiatri and the medical profession in antiquity", **Pap. Brit. School Rome** 45, 1977, pp.191-226; "Continuity and or rediscovery? The city physician in classical antiquity and mediaeval Italy", in A W Russell, **The Town and State Physician in Europe, from the Middle Ages to the Enlightenment**, Wolfenbüttel, Herzog August Bibliothek, 1981, pp. 11-21. My arguments on this point are not affected by the discovery at Ephesus, D Knibbe, "Neue Inschriften aus Ephesos, VIII", **Jahresh. ost. arch. Inst.** 53, 1981-2, pp. 136-140, n.136; cf **Zeit. Pap. Epigr.** 44, 1981, pp.1-10, of an inscription showing that tax-immunity to all doctors was granted by the triumvirs in the 40s BC (if not indeed by Caesar). It perhaps should also be emphasised that the legal position of Rome was so anomalous in this regard (see "Two notes on immunities", **Journ. Rom. Stud.** 61, 1971, pp. 52-63) that the organisation and rules for the college of doctors in Rome in 368 AD cannot with any degree of plausibility be transferred to the small towns of Asia Minor.

30. Galen, **De Medico Examinando** p. 193 Dietrich; XIV.621-3 K., with the commentary at **Corp. Med. Graec.** V.8.1., Berlin, Akademie Verlag, 1979, pp. 178-180.

31. F Kudlien, "Medical ethics and popular ethics in Greece and Rome", **Clio Medica** 5, 1970, pp. 91-121, esp. pp. 91-97; idem, "Zwei Interpretationen zum Hippokratischen Eid", **Gesnerus** 35, 1978, pp. 253-263; H M Koelbing, "Zu Fridolph Kudliens ´Zwei Interpretationen zum Hippokratischen Eid´", **Gesnerus** 36, 1979, pp. 155-157; R Wittern, "Die Unterlassung arztlicher Hilfeleistung in der griechischen Medizin der klassischen Zeit", **Münch. Med. Wochenschr.** 121, 1979, pp. 731-4.

32 Stobaeus, **Flor.** IV.38.9

33.   Particularly, **NH** XXIX.5.11 - 8.28

34.   H Gummerus, "Der Ärztestand in römischen Reiche", **Soc. Sci. Fennica, Comment. hist. et litt.** 3.6,   Helsinki,   1932;   Nutton, **The   Medical Profession**, p. 258.   The bias of the epigraphic sources towards the imperial **familia** may tilt the figures upwards, but the underlying pattern would still remain.   There is also a marked difference in civic profiles, as recorded epigraphically, between the doctors in the Greek East and those in the Latin West.

35.   See the list at Nutton, **The Medical Profession**, p. 262.

36.   Respectively, **IG** XIV.1934; **CIL** VI.9580 and **IG** XIV.1589, if there is a doctor; **IG** XIV.967; 1680;   1755 and 2014; 2019; an unpublished inscription in the Museo delle Terme to a Sosicrates Sosicratis f. Nicaésis medicus; L Robert, **Hellenica** 9,   Paris, 1950, p. 25; **Epigraph**.34, 1972, pp. 105-130, esp. pp. 126-130.

37.   A E Austin, **Cato the Censor**, Oxford, The Clarendon Press, 1978, pp. 157-181, 332-340.

38.   Pliny, **NH** XXIX.6.12f.

39.   His name is quoted by every writer on surgery in the renaissance I have seen, with varying emphases.   F Lefevre, **Les trois premiers livres de la chirurgie d´Hippocrate**, Paris, J Kerver, 1555, sig. c.1, calls him "Archibutus".

40.   Livy, **Periocha** XI; for an example of early Roman Hellenisation, cf. T J Cornell, "Aeneas and the twins: the development of the Roman foundation legend", **Proc. Cam. Phil. Soc.**, n.s. 21, 1975, pp. 1-32; cf. Phillips, art. cit., p. 268f.

41.   Dion. Halic. **Antiquit.** I.10, cf. X.53; Pliny´s own story implies the existence of other doctors.

42.   L Cohn-Haft, **The Public Physicians of Ancient Greece**, Northampton, Mass., 1956, p. 48, n.18.   There is thus no need to posit, with Baader, art. cit., p. 9, and Phillips, art. cit., p. 269, that Archagathus´ importance lies in his being "the first who tried to practise medicine scientifically" or the most distinguished practitioner heretofore. That might well be, but the Greek sources maintain a definite silence about his abilities, and the Peloponnese was not famous for its expert doctors. It is easy to see how Pliny could interpret Cassius Hemina´s account of the failure of the first city physician as the failure of the first doctor.

43.   Pliny, **NH** XXV.5.9-10;   XXVI.33.51; XXVI.6.11.

44. Cf. for local imitations of Laodicean nard ointment, Galen, X.791 K. The old survey of A Schmidt, **Drogen und Drogenhandel im Altertum**, Leipzig, A Barth, 1924, is still valuable.

45. J Scarborough, "Roman pharmacy and the eastern drug trade", **Pharmacy in History** 24, 1982, pp. 135-143. The views of J Innes Miller, **The Spice of the Roman Empire**, Oxford, The Clarendon Press, 1969, have been revealed as largely fanciful by the chalcenteric Manfred G Raschke, "New studies in Roman commerce with the east", in: H Temporini, ed., **Aufsteig und Niedergang der römischen Welt**, IX.2, Berlin, W de Gruyter, 1978, pp. 605-1361, esp. pp. 650-676. The sheer abundance of archaeological evidence presented by Raschke obscures the main lines of his argument, and the fact that he makes little use of the ancient writers on pharmacy.

46. Galen, XIV.30 K. Cf. also Pliny, **NH** XIII.2.17f.

47. Pliny, **NH** XXVI.1-6.9. It should be stressed that despite the lists of diseases assembled by Jones in the Loeb Pliny, vol. VII, pp. 547-553, very few of the diseases named in antiquity, particularly skin diseases and fevers, can be identified with anything like certainty.

48. Plutarch, **Table Talk**, VII.9; 732-733. Cf. also Seneca, **Ep.** 95; Maximus of Tyre, **Or.** 4.

49. Peter Garnsey, C R Whittaker, **Trade and Famine in Classical Antiquity**, Cambridge Philological Society, 1983, pp. 56-65.

50. Pliny, **NH** XXXIV.25.108, delivers a violent attack on the adulterations of the druggists: cf. also the evidence collected by Schmidt, op. cit., pp. 75-90. Other examples could be added from Christian sources.

51. The phrase is that of Polemo, Galen, XVIIA. 347 K. One can find hints of such an argument already in antiquity, in Pliny, **NH**, XIII.2.18, and, especially, in Gargilius Martialis 30; p.166 Rose. Cf. also J F Schulze, "Die Entwicklung der Medizin in Rom und das Verhältnis der Römer gegenüber der ärztlichen Tätigkeit von den Anfängen bis zum Beginn der Kaiserzeit", **Živa Antika** 21, 1971, pp. 485-505.

52. Especially **NH** XXIX.7.16-28; XXXIII. 57.164. Pliny´s price list is summarised by Schmidt, op. cit., pp. 104-6.

53. M Michler, **Das Spezialisierungsproblem und die antike Chirurgie**, Berne, Stuttgart, Vienna, Verlag Hans Huber, 1969.

54. Cf. L Edelstein, **Ancient Medicine**,

Baltimore, The Johns Hopkins Press, 1967, pp. 247-301; G E R Lloyd, **Science, Folklore and Ideology**, Cambridge University Press, 1983, pp. 149-167.

55. W D Smith, **The Hippocratic Tradition**, Ithaca and London, 1979, pp. 233-246.

56. L Edelstein, op. cit., pp. 173-191; E D Rawson, "The life and death of Asclepiades of Bithynia", **Class. Quart.** 32, 1982, pp. 358-370; J Pigeaud, "Sur le Méthodisme", **Mém. du Centre Jean Palerne** 3, 1982, pp.181-183.

57. A not entirely convincing defence of Thessalian methodism against Galen has been mounted by M Frede, in: J Barnes et al., **Science and Speculation**, Cambridge University Press, 1982, pp. 1-23. See also my article, "Language, style and context of the **Method of Healing**", in the Proceedings of the 2nd International Galen Conference, Kiel, 1982, forthcoming.

58. Galen, in particular, pokes fun at their claim to teach the whole of medicine within six months, but his strictures may apply to a type of Methodist long past, for his opinion of Soranus was far more favourable, and the writings of Soranus show a very competent practitioner; see Lloyd, op. cit., pp. 168-200.

59. The older article of Hans Diller, "Thessalos", **Real-Encyclopädie der classischen Altertumswissenschaft** II.11, 1936, cols. 168-182, remains fundamental.

60. I E Drabkin, "Soranus and his system of medicine", **Bull. Hist. Med.** 25, 1951 pp. 503-518; note also the warning of J Pigeaud, "Pro Caelio Aureliano", **Mém. Centre Jean Palerne** 3, 1982, pp.105-117.

61. Pliny, **NH** XXVI.8.14-17

62. J Ilberg, "A Cornelius Celsus und die Medizin in Rom", **Neue Jahrbücher** 19, 1907, pp. 377-412, is still worth consulting: see also P Mudry, **La préface du "De Medicina" de Celse**, Rome, Institu.- Suisse, 1982

63. Especially in **De Medicina**, VII-VIII.

64. Celsus, **De Med.** I. pr. 65.

65. Ibid., VII.26; cf. L C MacKinney, **Medical Illustrations in Mediaeval Manuscripts**, London, The Wellcome Historical Medical Library, 1965, p. 80f. and pl. 82a.

66. G Majno, **The Healing Hand**, Cambridge, Mass., Harvard University Press, 1975, pp. 186-188, 369f.

67. Professor J M Riddle, in a forthcoming

major study of Dioscorides, discusses the difficulties in determining the efficacy of ancient drugs. His suggestion to me, by letter, is that possibly as many as 40 percent of the drugs recommended for use in a particular disease would have had, if properly picked and stored, some positive action; some 5-10 percent would have had a negative action; and the rest would have been neutral or have no effect as yet confirmed by modern pharmacognosy. As well as this, one should not underestimate the placebo effect. Against this optimistic assessment, derived from the works of an excellent pharmacologist, must be set the great difficulties involved in securing top quality drugs, and in identifying plant species as well as individual diseases, and the possibility of wrong diagnosis. Furthermore, the growing fondness for polypharmacy would certainly have reduced the active qualities of each ingredient (for good and ill) within each preparation. Exact quantification is, of course, here impossible, but it might not be unfair to claim that in the learned pharmacy on the Greek model of Dioscorides, Celsus and Galen, 20 percent of drugs given would have a directly positive effect, while 5 percent or less would have a directly negative one, and that many patients would also have benefitted from the placebo effect. This is far from being the expensive inefficiency described by Pliny.

No comparable study in detail has been carried out on the folk remedies recommended by Cato and Pliny. Although, once again, the placebo effect would operate, my impression is that the range of actively efficacious drugs is considerably less, perhaps 5-10 percent.

68. On the theriac of Andromachus the elder, to which he gave the name "Tranquility", see Galen, **De Antidotis** 1.6: XIV.32-42 K. A better test is given by Ernst Heitsch, "Die griechische Dichterfragmente der römischen Kaiserzeit", **Abh. Akad. Wiss. Göttingen,** phil.- hist. Kl., 1964, n.62, pp. 17-25. The English survey by Gilbert Watson, **Theriac and Mithradatium,** London, The Wellcome Historical Medical Library, 1966, is very unreliable, cf. J Stannard, **Journ. Hist. Med.** 21, 1966, p. 430. There is much of value in Thomas Holste, **Der Theriakkrämer**, Pattensen, Wellm Verlag, 1976.

69. Cf. P Brain, **Galen´s Pathology,** Diss., Durban, 1982, pp.93-123, and the later arguments summarised by I Lonie, "Fever pathology in the

sixteenth century: tradition and innovation", in: W F Bynum, V Nutton, **Theories of Fever from Antiquity to the Enlightenment,** London, Wellcome Institute for the History of Medicine, 1981, pp. 19-44.

70. The identification of ancient skin diseases was already regarded as problematical in the late fifteenth century: see N Leoniceno, **De Morbo Gallico,** Venice, Aldus Manutius, 1497, attempting to distinguish between the skin diseases recorded by Galen, Celsus and Avicenna. His example was often followed in the next century. For the problem of **lepra** and leprosy, see Julius Preuss, **Biblical and Talmudic Medicine,** New York, London, Sanhedrin Press, 1978 (originally 1911), pp. 323-341; Stanley G Browne, **Leprosy in the Bible,** 3rd ed., London, Christian Medical Fellowship Publications, 1979.

71. On the difficulty of reconciling ancient and modern diagnoses in one instance, see my note on a case of Galen, **On Prognosis,** 8, at **Corp. Med. Graec.** V.8.1, pp. 203-208. On the adulteration of drugs, Schmidt, **Drogen,** pp. 114-126, is enlightening.

72. L Edelstein, op. cit., pp. 328-331.

73. K Deichgräber, "Professio Medici. Zum Vorwort des Scribonius Largus", **Abh. Akad. Mainz,** Geistes- u. Sozialwiss. Kl., 1950.9.

74. Galen, XVIIA.150f. K.; XVIIB.147 K; 222-233 K. Cf. also K Deichgräber, "Medicus gratiosus", **Abh. Akad. Mainz,** Geistes- u. Sozialwiss. Kl. 1970.3.

75. I E Drabkin, "Soranus and his system of medicine", pp. 503-518.

76. See in particular, P M Green, **Prolegomena to the Study of Magic and Superstition in the elder Pliny,** Fellowship Diss., Cambridge, 1954

77. Pliny, **NH** XXX.1.1-6.18; Deichgräber, **Professio Medici.**

78. Pliny, **NH,** XIV.1.3f.

79. On the habits of the Palaeothebans, see Plutarch, **Table Talk** VII.1; on the Marsi, see my article "The drug trade in antiquity", **Journal of the Royal Society of Medicine** 78, 1985, 138 - 145.

80. Cicero, **Pro Cluentio** 40. See also F Kudlien, "Schaustellerei und Heilmittelvertrieb in der Antike", **Gesnerus** 40, 1983, pp.91-98.

81. Dio Chrys., **Or.** 9.4.; cf. also, as well as St Luke, Galen, V.18.K.; XI.357 K.; Plinius Iunior, **De Medicina,** prol 1.

82. Pliny, **NH** XVI.18.40: XXXIV.11.108; Valerius Maximus, IX.1.

83.   G E R Lloyd, **Science, Folklore and Ideology,** pp. 135-149.

84.   Anon. Lond. XXIV.30; Pliny, **NH** XXVI.8.14; cf. XIV.17.96, and **CIL** x.388, the ´wine-giving´ doctor, Q Manneius of Tralles, and later of Atinum.

85.   J Wight Duff, **A Literary History of Rome,** 3rd ed., London, E Benn, 1964, p. 308.

Chapter Five

PHARMACY IN PLINY´S **NATURAL HISTORY:**
SOME OBSERVATIONS ON SUBSTANCES AND SOURCES

J Scarborough

Very prominent in books XX through to XXXII of the
Natural History are drugs and medicinals that appear
to span the entire scope of pharmacy of classical
antiquity. One is regaled with presumed quotations
of sources, from the quasi-mystical Magi to
straightforward writers who include Celsus and
Sextius Niger, as well as obvious magical remedies
that are part of a hoary folk tradition not limited
to Italy, but which includes matter from around the
mediterranean dominions of Rome. Moreover, if the
reader performs the arduous task of counting
varieties of plants, minerals, animal drugs and
related substances employed as drugs - omitting the
synonyms and repetitions - he emerges with an
astounding array of over 900 substances, well over
the nearly 600 remedies described by Dioscorides,
the almost 550 species of plants contained in the
works of Theophrastus, and even in excess of the 650
or so remedies assembled by Galen and the over 500
drugs listed by Paul of Aegina. The very bulk of
Pliny´s data defies simple analysis, and it is clear
enough that Pliny is quite proud of his collection
of details on drugs and drug lore, as if he did not
expect a reader to question either his sources or
his substances: all were part of the natural world
and its multitudinous wonders. This enthusiastic,
breathless mood has endeared Pliny to centuries of
readers, and the tumbling of facts, tales, stories
told for amusement, anecdotes, legends and ´things
heard´ carries us along as a good story should.
One soon recognises, however, that embedded in this
huge assemblage are underlying sources for the
history of Greek and Roman pharmacy, now generally
lost, and that untangling those sources would
provide great insight into the development of
pharmacy before Pliny´s day.

In the beginning of book XX Pliny tells us why he
has decided to include medicinal plants in his
gigantic pot-pourri of natural wonders and facts
gleaned from voluminous reading and wide personal
experience, and he begins where a Roman thought he
should: in his garden or farm.

> **Maximum hinc opus naturae ordiemur et cibos**
> **suos homini narrabimus faterique cogemus ignota**
> **esse per quae vivat. Nemo id parvum ac**
> **modicum existimaverit nominum vilitate**
> **deceptus.**
> "Now we begin the greatest of Nature´s works
> and we will address foods for man, and compel
> him to admit that he is ignorant about how he
> lives. Let no one, deceived by the common
> quality of the names, think that this is of
> little value or of moderate importance."
> (**Natural History** XX.1)

This initial statement by Pliny about the
forthcomning consideration of medicinals and healing
substances is of crucial importance in understanding
Roman concepts concerning pharmaceuticals, as well
as in explaining why Pliny would speak of "foods"
here, when he has given extensive coverage to
cereals, the preparation of bread, garden vegetables
and the like in books XVIII and XIX: drugs are to be
understood as an integral part of the lore of food,
and thereby an important emphasis throughout Pliny´s
thinking about pharmacy would be in terms of medical
dietetics. Foods and medicines are to be, at once,
sources of wonder and admiration, and Pliny´s
avowed devotion to the "old Roman ideal"(1) will
lead him repeatedly to assert how plants - even the
most common - can provide mankind with healing
drugs, particularly those plants a Roman farmer or
citizen of the numerous Italian towns would know
from childhood.(2) This point needs to be stressed
at the beginning of a brief overview of Pliny´s drug
lore and the sources for that knowledge as they
appear in the **Natural History**: in contrast to the
twentieth-century ignorance of plants in the field,
most of humanity in classical antiquity dwelled on
farms or lived very close to the life of a
non-urbanised countryside. Pliny´s account of
drugs is bedecked with multitudinous quotations from
"foreign authorities", and he is indeed fascinated
by what he has found in the Greek treatises on
medicine and pharmacy, but he also infuses these
data with a native Italian knowledge of plants and

herbs.(3)   We may rightly call much of this
´folklore´ and many of the suggested remedies are in
the purely magical or folkloristic class, but Pliny
consciously separates them from his "Greek authors"
as he culls information from all sources.(4)
Noteworthy among his **auctores** - Roman authors - of
works on drugs and medicinals are tracts by Sextius
Niger(5) and Julius Bassus,(6) both of whom wrote in
Greek.   The context, however, of almost all the
drugs in Pliny´s **Natural History** is one of the
fullness of Roman Italy, and this setting lends a
certain omnivorous cast to his writing, which is the
result of his assumption that **all** aspects of human
investigation of nature - from the completely
magical to the absolutely philosophical - are fit
subjects for his encyclopedia.(7)
    In some respects, Pliny is writing for an
´educated public´ in the middle of the first
century, a public that would be curious about odd
tales and local customs, peculiar animals of legend
and Greek biology as well as what the ´latest
thinking´ might be about Italian flora and fauna,
and Pliny could assume a readership that would be as
intelligent and curious as he was.(8)   The **Natural
History** met a need filled today by a comprehensive
popular encyclopedia, and the pharmaceutical
sections served the same purpose as seen today in
the revival of books on herbal lore.   One is struck
at first by Pliny´s general lack of interest in
diseases themselves, although there is ample
citation of various names, gradations, and technical
terms which can be grouped and classed according to
Greco-Roman pathology;(9) but Pliny is far more
interested in any drugs or remedies suggested for
their cure, and one calls to mind similar approaches
in modern handbooks of herbs, as by Grieve, Flück,
and Schauenberg and Paris.(10)   This does not make
herbal lore any less useful or valid in its ancient
Roman or modern versions,(11) but this emphasis on
cure rather than disease (or perhaps more strictly
medical theory) indicates the difference between a
´formal medicine´ and a more encompassing medicine
´practised´ by others than physicians.(12)
    There is little doubt that Pliny is well aware of
Greek medical writers, from Diocles and the
Hippocratics to the Greek physicians of his own day,
and many of the ´medical´ books in the **Natural
History** are taken up with drugs and pharmaceuticals,
drawn from a presumed large variety of sources in
both Greek and Latin.(13)   Yet there are many
curious errors in the ´facts and observations´

about drugs, and Pliny´s rampant and childlike curiosity led to a too rapid compilation of pharmaceutical data, shown by a lack of attention to specific detail and an apparent lack in cross-checking references and the omission of revisions. Some of these mistakes can be corrected through parallel readings in the Greek of Dioscorides, who used common sources cited by Pliny, particularly Sextius Niger´s work on medicinal plants,(14) written in Greek.

Books XX-XXXIV of the **Natural History** are stuffed with herbal and medical lore, and Pliny´s own listing of the raw numbers of details is in itself staggering: **medicinae** together with **historiae** and **observationes** in the fourteen books comes to a total (as I have added them) of 13,805 ´facts´.(15) Not all of these data are pharmaceutical, but a good majority are concerned with remedies, purported powers and quoted sources on drug lore, so that Pliny is faced with an enormous task of classification. How would he draw together all these varied bits into a comprehensive and useful whole? Other medical writers wrestled with this problem, and in Latin, Celsus´ methods of classification of drugs are probably the most streamlined in classical antiquity.(16) Yet there were as many ways of listing drugs as there were authors who compiled such remedies: the Hippocratic writers generally listed the diseases and then the drugs useful in treating those ailments;(17) Theophrastos left his herbal lore until book IX of the **Historia Plantarum** and attempted a poorly defined taxonomy of herbs while relying on the observations of the everyday **rhizotomoi**, who did not consider taxonomy as much as effects when a given drug was administered;(18) the pseudo-Aristotelian **Problems** seems to have attempted to class drugs according to their essential qualities, with an equally imprecise result;(19) Apollodorus chose toxic effect for his drug classifications,(20) and the obtuse poems of Nicander record the muddled nature of this attempt;(21) and Dioscorides - a contemporary of Pliny, but apparently unknown to him - invented a brilliant ´drugs-by-affinity´ system,(22) which was to be completely ignored by pharmacologists in later centuries, including the polymatic and chalcenteric Galen.

It is curious that Pliny did not know the work of Scribonius Largus, whose **Conpositiones** was set down during the reign of Claudius, and the contrast between Scribonius´ drug lore and that of Pliny is instructive. Scribonius, as noted by Rinne,(23) unites book-learning with experienced, practical knowledge of drugs, so that the **Conpositiones** records what a ´professional druggist´ would do in classification of medicinals:(24) one does get a kind of rough head-to-toe arrangement, but crucial to Scribonius is the prepared form of his drugs. Decoctions, infusions, plasters, salves, ointments, potions ´and so on provide the basic framework. Thus each of the 242 plants, 36 minerals, 27 animal products and eight poisons have specific methods of preparation in the **Conpositiones** - similar in many respects to a large portion of Galen´s huge compilation of drugs and drug recipes(25) - whereas Pliny´s drug lore has only a very loose classification by types of plants that yielded drugs, for example, book XX of the **Natural History** with **medicinae ex his quae in hortis seruntur**, XXI with **naturae florum et coronamentorum** (with drugs derived from the rose beginning at XXI.13), XXII as **continentur auctoritas herbarum** (as if Pliny were culling without worrying about classing his herbs, except that many in XXII are wreath-flowers, trailing in from XXI), XXIII **medicinae ex arboribus cultis**, and XXIV **medicinae ex arboribus silvestribus** (six drugs here (XXIV.10) from **loto Italica**) and so on. Thus Pliny´s drug lore is a medley of facts and details, taken both from his written sources and from current oral traditions, meshed with inaccuracies resulting from rapid reading and recording and an occasional blind eye to the ridiculous in his sources. One can of course simply argue that Pliny was giving an accurate account of such sources, but the self-contradictions suggest that he either did not notice variations in his written materials, or that his purpose was **not** to compile a well digested encyclopedia (as was the case with Celsus´ complete work) but a series of notes on intriguing aspects of nature that would be an inclusive record of knowledge of natural history as he had found it in his own time. Stannard has argued succinctly that Pliny´s command of botany, at least for Italian flora, is reasonably good,(26) and one may also suggest that Pliny´s acumen in recording his Greek sources is often rather high.(27) There are, however, numerous problems that exist in the texts of the **Natural History** that contain medical botany and pharmacology, and one has to approach the question of Pliny´s texts on two levels: his sources; and what are the actual species and substances within the huge gathering of materials on drug lore.

Some Sources for Pliny´s Drug Lore
One of the odd facts of early imperial Roman
pharmacy is that both Dioscorides and Pliny
published their respective works within two decades
of one another, and neither seems aware of the
other. Textual and historical evidence shows that
Dioscorides´ **Materia Medica** appeared about 64 AD,
and Pliny´s **Natural History** is firmly dated 77 AD.
Thus any correspondence of specific description or
details between Pliny´s drugs and those of
Dioscorides emerges from both authors having
independently employed common sources. This
enables the modern scholar to make judicious
comparisons between the two sets of quoted
materials, and such an analysis becomes doubly
valuable since it allows us to recover some sense of
the contents of earlier works on pharmacy and
pharmacology, which have disappeared except in
quotations in Dioscorides, Pliny, Galen, Oribasius
and other Roman and early Byzantine writers. The
first document of importance in a comparative
examination is the difficult **Preface** by Dioscorides
to his **Materia Medica**.(28)
    Unlike Pliny, Dioscorides was a physician -
perhaps even an army doctor,(29) although our
evidence is far from secure - with a special
interest in medical botany and general pharmacology.
Recent research has shown that Dioscorides most
likely began his study of plants and medical botany
at Tarsus,(30) which apparently functioned as a
famous ´teaching centre´ for pharmacy. Various
other and previously obscure names are now linked
with this Tarsian tradition that reaches back at
least several generations before Dioscorides´ own
time.(31) It is clear also that Dioscorides has
read deeply in the medical and pharmacological
literature available to him in Greek, and even
though the well known names of Hippocrates,
Theophrastos and Nicander only appear rarely as
cited authorities in the **Materia Medica**,(32) there
is an indication that data from these three very
famous writers have made their way into Dioscorides´
text. It is, however, in the short - but
exceedingly difficult - **Preface** that Dioscorides
tells us which authorities are before him, and why
he believes that his work on medical botany and
general pharmacology will be much better than all of
the previous authorities: "...some of my

predecessors did not give a complete survey, while others took most of their information from written sources".(33) Then he begins to name these predecessors:

> ...Iollas of Bithynia and Heraclides of Tarentum touched on only a small portion of the subject, entirely omitted the botanical tradition, and made no mention at all of metallic drugs and spices. Crateuas the Rootcutter and Andreas the Physician - who are apparently more precise in this aspect - omitted many exceptionally useful roots and a few herbs.
> However, one must admit that the older authors combined the paucity of their information with precision, in contrast to recent writers like Julius Bassus, Niceratus, Petronius, Niger, and Diodotus, all followers of Asclepiades.(34)

Areius, the individual to whom the **Materia Medica** is addressed, was a doctor and a pharmacologist in his own right,(35) but is not quoted directly by Dioscorides. Other authorities cited in the **Materia Medica** include Diagoras (IV.64), Erasistratus (IV.64), Iobas (III.82), Mnesidemus (IV.64) and Philonides (IV.148) and perhaps Thessalus (I.26) and Peteesius (V.98). This list of 19 sources is not necessarily complete, but does provide a convenient group of names which can be compared with the far less specific listings of sources given by Pliny. From Dioscorides we can learn the names of authors of works particularly devoted to pharmacy, which possibly can be correlated with the names and quotations in Pliny´s **Natural History**; then a comparison can be made of the quoted extracts in terms of some of the specific pharmaceuticals that are mentioned.

There are Greek sources used by Pliny which are not cited by Dioscorides, so that one must also consider quotations by Pliny from Apollodorus (as opposed to Apollodorus through Nicander), Aristophanes, Menander, Callimachus, Diocles and Xenocrates. After detailed consideration, it seems probable that Pliny employed one set of authorities for strictly pharmacological information (Sextius Niger, Erasistratus, Iollas, Diagoras, Mnesidemus (that is, Mnesides in Pliny), Heraclides of Tarentum and the Asclepiadeans Julius Bassus, Niceratus, Petronius, Diodotus and Philonides), and a second

group of authors and works for the names of plants
(the Hippocratic corpus, Diocles, Apollodorus and
the poetry of Aristophanes, Menander, Callimachus
and Nicander); the doxographical evidence, best seen
in Athenaeus, shows a rich history of plant names in
the "old Greek" poets.(36)   A third kind of written
authority, used by Pliny as an important part of his
basic information on medical botany, would be
represented by Xenocrates, whose **Peri tēs apo tōn
phytōn ōphelias (Useful things from plants)**was
based, in turn, on a number of occult and
quasi-occult authors and works that included the
Physika by Democritus Bolus, the "Magi" (Ostanes and
Zoroaster), Agathocles′ **Peri Diaites (Regimen)**, and
various Pythagorean and neo-Pythagorean writers.(37)
If one compiles further citations of authorities,
as given by Pliny for his botanical, medical and
pharmaceutical books, there results a lengthy
listing.   There is, however, something rather
misleading about these long lists of authorities,
which can be illustrated by a consideration of
Pliny′s handling of garden vegetables and herbs.(38)
In addition to Sextius Niger and Xenocrates, there
seems to be a third Greek source used by Pliny,
indicated by a uniformity of repeated details from
varied sources, a uniformity that corresponds to the
quotations from Sextius Niger and Xenocrates.   One
may immediately suspect some kind of doxographical
collection that would, indeed, include extracts from
the Hippocratic corpus,(39) as well as the
following: Philistion, Diocles, Chrysippus,
Praxagoras, Pleistonicus, Dieuches, Medius, Simus,
Cleophantus, Pythagoras, Glaucias, Dionysius,
Apollodorus, Diodorus, Heraclides of Tarentum,
Chrysermus, the followers of Erasistratus(40) as
contrasted to the followers of Asclepiades,(41)
Solon of Smyrna and probably Sosimenes and
Tlepolemus.   One could add the name of
Theodorus(42) to this doxographical collection, but
he is identified among the pneumatics by Athenaeus,
so that Pliny gained this source through Xenocrates,
not in the collection of extracts represented by the
listing of previous names.**(43)**
Once these "names" have been linked to their
respective passages in the **Natural History**, it is
apparent that Pliny has consulted a collection of
writings that summarised a medical botany that
emphasised dietetics: Chrysippus,(44) Democritus
joined to Diocles and Dionysius,(45)
Democritus,(46) Hicesius(47) and Erasistratus.(48)
In the **Natural History** XX.187-190, Pliny employs a

characteristic phraseology(49) that links with the
sunt qui in the closing section of XX.195, which
smooths out the quotations - so it seems at first
glance - from Evenor and Iollas.(50)   Textual
parallels, however, in Dioscorides, Galen, Symeon
Seth(51) and others(52) show that the full block of
material has come from Sextius Niger.
Tlepolemus, along with Dalion, Sosimenes, Dieuches,
Evenor and Heraclides of Tarentum are all cited
alongside Sextius Niger,(53) but a section in
Dioscorides shows that the information on the anise
had been compacted by Niger, even though Pliny has
recorded the names of the multiple authors, while
Dioscorides has not.(54)

It appears, therefore, that Sextius Niger has in
his turn consulted and extracted a collection of
pharmacological opinions, particularly emphasising
the medical botany of dietetics.   Wellmann has
proposed that this doxographical collection came
from Hellenistic Smyrna,(55) which had a long
tradition of medical figures who studied pharmacy
and dietetics.   The most likely candidate as the
author of the Smyrna collection, which was the basis
for Niger´s account, is Solon of Smyrna, known
through Galen as the "dietician".(56)   The key in
identifying this doxographical source underneath
Sextius Niger as employed in the **Natural History** is
Pliny´s **sunt qui**,(57) and Solon appears clearly as a
mingled part of a doxographical tradition.(58)   As
one reads **Natural History** XX.235,(59) it appears
reasonably certain that Solon of Smyrna is the
author of this third Greek source, heavily employed
by Pliny in his herbal medicines in books XX through
to XXVII.   Animal drugs, that is, pharmaceuticals
that are derived from animals,(60) come in large
part from another work by Xenocrates of Cilician
Aphrodisias(61) titled **Peri tēs apo tou Anthrōpou
kai tōn Zōōn ōpheleias (Useful (things) for Man from
Animals),** as well as from the **Physika** of pseudo-
Democritus.(62)   Apollodorus and Nicander do appear
in the listings of authorities for books XXVII -
XXX, Nicander for XXXI -XXXII,(63) and the
mysterious Licinius Macer for XXXII.   Sextius Niger
is listed for books XXVIII - XXX and XXXII.

Latin sources for the drug books in the **Natural
History** are less troublesome.   A major source of
data for Pliny is the lost work by Valgius Rufus, as
well as a tract by Antonius Castor,(64) but it seems
clear that Pliny has excerpted much of his Latin
medical botany from the expected writings available
in his own time: Cato,(65) Varro and Celsus.   One

also finds quotations from an Asclepiades,(66) and
these data are generally from the drugbooks of
Asclepiades of Bithynia,(67) perhaps ocasionally
conflated with the pharmacy of Asclepiades
Pharmacion, known through fairly copious citations
in Galen´s summary of drug lore.(68)   Pliny
borrowed heavily from Celsus´ botanical and herbal
lore,(69) but one can also make the argument that
Pliny did, indeed, have a text of Theophrastos
before him as well.(70)   One cannot, however,
escape the impression that Pliny´s main sources for
his pharmaceutical lore were either encyclopedic
(for example Celsus) or doxographical (Solon of
Smyrna, for example), with certain, nearly
contemporary works that supplied not only masses of
data, but also the names of quoted authorities,
which would be duly recorded by Pliny amongst **his**
sources.   Sextius Niger, Xenocrates and Solon of
Smyrna emerge as Pliny´s major references for many
of the facts gathered from the Greek, and the
traditional Latin lore assembled by Cato, Varro(71)
and perhaps Valgius Rufus formed a core for Italian
materials.   Celsus,(72) of course, would give Pliny
a conduit into both a native Latin tradition
coupled with a shrewdly perceived Hellenistic drug
lore, so that Celsus´ **De Medicina** looms as a far
more important source for Pliny´s pharmaceuticals
than has been previously assumed.

Given the evidence which indicates that Pliny
most often used written sources of the first 75
years of the Roman Empire, one can assume that Solon
of Smyrna´s compilation was in circulation some time
after Celsus published his encyclopedia (during the
reign of Tiberius), and that the works of Sextius
Niger and Xenocrates were very nearly contemporary
with Pliny himself.   Xenocrates, however, is
something of a problem: Pliny´s citations reveal
either a confusion between an ´Elder´ and a
´Younger´, or his employment of some care in an
attempt to distinguish between the ´stone books´ of
Xenocrates of Ephesus and the ´medical books´ of
Xenocrates of Aphrodisias.   Fabricius(73) represents the current thinking among specialist
scholars that Xenocrates of Aphrodisias lived in the
second half of the first century, but Ullmann
follows the earlier suggestion of Wellman that
Xenocrates of Ephesus also lived during the reign of
Nero,(74) and both would be contemporary with Pliny.
If we believe the opinions of Galen, Xenocrates of
Aphrodisias was something of a charlatan, even as he
assembled a fat collection of quoted sources

on drugs made from animals and animal parts. The firm dating of Fabricius and Kudlien seems supported by Galen´s statement that Xenocrates was alive "...during the time of our great-grandfathers",(75) putting Xenocrates of Aphrodisias´ **floruit** at about 50 AD. It is not impossible, as observed by Ullmann, that there might be two authors named Xenocrates (an ´Elder´ from Ephesus and a ´Younger´ from Aphrodisias), but it only adds to the complexity to assume two separate authorities called Xenocrates, both of whom wrote on medical topics at roughly the same time. Given the long fragment from Xenocrates´ **Food from Water Animals**, preserved by Oribasius,(76) and the notices by Galen of a tract called **Useful (Things, viz. Drugs) from Animals**,(77) as well as the ´medical nature´ of Xenocrates´ **Book on Stones**,(78) it appears far more reasonable to suppose that **a** Xenocrates was the author of a number of works, cited as part of a pharmaco-doxographic tradition by Pliny. Perhaps there has been confusion by various scribes in the textual history between "Ephesus" and "Aphrodisias", which might have been insignificant especially to copyists in the Latin West. Xenocrates´ writings were very important to Pliny, as the numerous citations suggest, and the fragments of the original Greek treatises show how the state of medical dietetics and medical zoology - and their relationship to pharmacology - had developed in Greek works on the topic during the first half of the first century. It may be noted, in passing, that Xenocrates apparently represented a long-lived tradition in medical dietetics that can be traced back to Aristotelian thinkers and Hippocratic physicians, and which was to claim much attention in the Hellenistic era, as seen in the remnants of the writings of Diphilus of Siphnos.(79)

Sources and Substances in Pliny´s Drug Lore
Even though Dioscorides does not often name specific written sources in his **Materia Medica**, there are numerous parallels in the **Materia Medica** with passages in Pliny´s **Natural History** which do bear the names of cited authorities. Since Solon of Smyrna probably underlies Sextius Niger´s account, and would therefore provide many of the ´names´ in Pliny´s extensive listing, one can conclude that the parallel passages in Pliny and Dioscorides are derived independently from Sextius Niger. Dioscorides has rejected the simple acceptance of

his written sources on pharmacology, in favour of seeking the plants and testing them as drugs,(80) but there is clear evidence that he did, indeed, record data on drugs from his written sources, especially Sextius Niger. Thanks to Dioscorides' care and precision, particularly as he confirms or denies statements found in his sources, we can gain insight into the way in which Pliny used his written materials to gain specifics on pharmaceuticals. It was very unlikely that direct use was made by Dioscorides of those he names as "recent writers",(81) except for Niger, given his ridicule for such sources as "...vain prating about causation".(82) Dioscorides shows that of all his sources, he had studied the work of Sextius Niger most closely, and says the following about his botanical, medical and compiling abilities:

> ...Niger, who seems to be the best (of the recent writers) says that spurge-resin is the juice from the olive-spurge (**Daphne oleoides** Schreb.) and that perfoliated St John's wort (**Hypericum perfoliatum** L.) is the same as triangular St John's wort (**Hypericum triquetrifolium** Turra); he says that bitter aloe (**Aloe perryi** Baker) is dug up in Judea and makes many equally wrong statements in defiance of manifest truth, which proves that he took his evidence not from his own eyes but from faulty secondhand written sources. Niger and the rest also made mistakes in the organization of their material, some throwing together incompatible properties, others using an alphabetic arrangement which splits off genera and properties from what most resembles them.(83)

These lines from the **Preface** (3) of Dioscorides' **Materia Medica** show several characteristics of Sextius Niger's writing on medical botany and pharmacy: he had based his account on earlier treatises, an aspect of Niger's writing discerned in Pliny's use of Niger and the underpinnings of Solon of Smyrna; Niger had compiled his works on drugs on an alphabetic arrangement, also seen in some of Pliny's excerpts;(84) Niger had not sought the plants in the field, but relied on his written sources, which led to gross errors cited here (confusing the prepared form of aloe with a mineral),(85) confusing spurge-resin (derived from a North African shrub, **Euphorbia resinifera** Berg.)(86)

with the olive-spurge; and mauling the descriptions of the two different kinds of St John´s wort,(87) and dependence on a quasi-philosophical doctrine of properties ("Asclepiadean") that is also occasionally mirrored in Pliny.(88)

As contrasted to the murky state of our knowledge of the texts of Solon of Smyrna and Xenocrates, we can make some fairly assured judgements about the writings of Sextius Niger. Dioscorides has singled him out for major criticism, indicating that it was Niger´s "drug books" - compiled from many earlier works - that Dioscorides knew best of his predecessors, and the close approximation in many passages of the **Materia Medica** and the **Natural History** delineate how a skilled field pharmacognosist would employ the facts of Niger´s compilation, as compared with an encyclopedist making a new compilation based on older ones. Sextius Niger as quoted and used by both Pliny and Dioscorides provides us with specific substances, and a few selected corollary sections in the **Materia Medica** and **Natural History** will enable us to see how Pliny made use of specific details in one of his sources.

In his consideration of a plant called **epimedion**, Pliny (XXVII.76) says that it has a stem of moderate size and ten or twelve leaves that resemble those of the ivy. Moreover, he continues, the plant never bears flowers, has a thin black root, a heavy odour, a feeble taste and grows in humid soils.

> **Et huic spissandi refrigerandique natura, feminis cavenda. Folia in vino trita virginum mammas cohibent.**

> And this plant has thickening and cooling properties, and it should be avoided by women. Beaten up in wine, the leaves restrain the growth of the breasts in young women. (89)

The parallel passages on the **epimedion** in Dioscorides, **Materia Medica**, IV.19 would read in translation:

> **Epimedion:** it has a small stalk, ten or twelve leaves resembling ivy; it bears neither seed nor flower; the roots are thin, black, strong-smelling, and taste flat; it grows in wet places.
> The leaves pounded up with oil make a plaster for the breasts, applied so that they do not grow large. The root is contraceptive

71

(**atokios**), and the leaves when pounded up in
wine and drunk after the menstrual flow (five
drachams for five days) prevent conception).(90)

The common source for the two accounts is Sextius
Niger, but one looks with fascination at what Pliny
has omitted from his text: the description of the
contraceptive use of the plant. Pliny has also
confused the "pounding with oil" of Niger (the
breast plaster) with the "pounding in wine" (the
post-menstrual contraceptive). The "thickening and
cooling properties" is an addition by Pliny, which
may have been inserted from one of his Hippocratic
texts, but generally (except for the omission of the
contraceptive prescription) Pliny has been quite
faithful to his Greek original. The Latin text may
bear the marks of scribal tampering, and the **feminis
cavenda** could easily be an intentional interpolation
by uneasy medieval copyists, since the rest of the
passage reflects rather well what Dioscorides has
also recorded in Greek.

This is one of those cases where Dioscorides has
not verified the plant, since the description is
poor, no matter how one interprets the Greek.
Identification of **epimedion** is very uncertain,(91)
but Pamphilus (Dioscorides, IV.19 RV) gives a series
of synonyms for the plant that includes **erinos**,
which would make it some kind of basil-like herb,
shown by Nicander, Diocles and elsewhere in
Pliny.(92) If **epimedion** is equivalent to **erinos**,
then it is possibly **Campanula rapunculoides** L., the
creeping bellflower, very poorly described. If
Greco-Roman pharmacy had indeed prescribed
contraceptive powers for **epimedion** as the supposed
equivalent of **erinos**, such have dropped out of the
descriptions recorded by Paul of Aegina.(93) It
seems that ancient botanists did not observe the
bellflower in July and August, the normal time for
the blooming of its purple flowers. Pliny does not
know the plant, and merely copies from his source,
Sextius Niger.

Sextius Niger is the common source for Pliny,
Natural History XXII.51-52, and Dioscorides, **Materia
Medica** IV.24-25. Comparison of these two extant
accounts can suggest further points in regard to
Pliny´s acquaintance with specific drugs and his
employment of written sources. In translation,
Pliny, XXII.51-52:

There is another plant also, properly called

onochilon, named by some anchusa, by others archebion, by still others onochelis or rhexia, and most commonly enchrysa. It has a small stalk, a purple flower, rough leaves and branches, a root the colour of blood at harvest time, though black at other times, and it grows in sandy soils. It is an antidote for snakebite, especially poisonous snakes, both the root and the leaves being equally efficacious in food and drink. It has its properties (**vires**) at harvest time. When pounded, its leaves smell like cucumber. (52) It is given in doses of three cyathi for a prolapsed uterus. With marjoram (**hysopum**) it expels tapeworms, and for pains in the kidneys or liver, it is drunk in hydromel (**aqua mulsa**) if one has a fever, but otherwise, in wine. The root is smeared on freckles and psoriatic sores (**lentigini ac lepris radix inlinitur**). It is said that those who carry the root are not bitten by snakes. And there is another plant similar to this one, but smaller, with a red flower, and it is used in the same way as the other one (above); and they record that if it is chewed thoroughly and spat out on to a snake, it dies. (94)

In translation, Dioscorides, IV.25-25:

Another alkanet (**anchousa**), which some call Alkibiadeion or onocheiles. This differs from the previous one in that it has smaller leaves, but they are sharp like those of the previous plant; it has thin branches, on which is a flower that is reddish-purple, tending towards Phoenician purple; it has very long red roots, at harvest time having (in them) something like blood; it grows in sandy places.

Its property (**dynamis**) and of the leaves is to be an antidote for those bitten by wild animals, especially for those bitten by poisonous snakes (**echiodektois**); (the root and leaves) are eaten, or drunk, or hung about one´s neck. And if someone has chewed it well, he should spit it out into the mouth of the wild animal to kill it.

(25) And there is another plant, resembling the previous one, but having a smaller seed the colour of Phoenician purple: if someone chews up the seed thoroughly and spits it into the mouth of the snake (**herpeton**) it dies. The root

73

being drunk with marjoram (**hyssopos**) and garden
cress (**kardamon**) in the amount of an **oxybaphon**
expels the tapeworm.(95)

The "other alkanet" is probably **Echium diffusum**
Sibth.,(96) a bugloss in the Boraginaceae, and the
"smaller one" may be **Lithospermum fruticosum** L.,(97)
or **L. diffusam** Lag., both being gromwells also in
the Boraginaceae. In this case Pliny's framework
is a text of Sextius Niger, but he has added bits
from other sources on the crushed leaves smelling
like cucumbers and on the use of the root as a
liniment (an embrocation usually made with oil) to
smear on freckles and scaly and red skin sores.
Pliny has received the folk tradition of spitting
the masticated leaves into the "snake's mouth" to
kill it, and he has modified it into the more
'sensible' spitting the chewed leaves **on** the snake.
Pliny has retained the marjoram (**Origanum vulgare**
L.) to expel tapeworms, but he has omitted the
garden cress (**Lepidum sativum** L.) in Niger's
prescription. Both plants are fairly common in
Italy, so that Pliny's omission has no real
explanation, unless this is the result of a rapid
skimming of the Greek text. In summary, in **Natural
History**, XXII.51-52, Pliny has been fairly faithful
to his Greek source with the exception of the garden
cress.

Very similar employment of Sextius Niger by Pliny
can be documented in numerous instances paralleled
in Dioscorides,(98) and each is suggestive of the
manner in which Pliny used his pharmaco-
doxographical sources. The following are merely
illustrative. **Natural History**, XXVII.74 (**elatine**)
compared with Dioscorides, IV.40 (**elatine**) shows
that Pliny has used a text by Sextius Niger, who is
probably quoting Crateuas, and the plant is probably
Kickxia spuria (L.) Dumort, the round-leaved
fluellen,(99) used for 'rheumy eyes' and diarrhoea.
Natural History, XXVII.93 (**Ideae herbae**) compared
with Dioscorides, IV.44 (**idaia rhiza**) again
indicates that Pliny was using a text by Sextius
Niger, probably quoting Crateuas, and the herb is
most likely **Streptopus amplexifolius** (L.) DC, a
mountain broom, used for loose bowels and for a
woman's excessive menstruation. **Natural History**,
XXXXVI.132 (**Equisaetum hippuris Graecis dicta**)
compared with Dioscorides, IV.46.1 (**hippouris**) again
reveals that Pliny is quoting from a Greek text by
Sextius Niger, quoting Crateuas, and the specific
plant under consideration is one of the **Equisetum**

species (horsetails, here probably **E. arvense** L., or perhaps **E. limosum** L. or **E.palustre** L.), used as a diuretic and as a styptic; but Pliny in **Natural History** XXVI.133 has switched to another source, and then returned to Sextius Niger at XXVI.134, as comparison with Dioscorides, IV.46.2 and 47 would suggest.   Pliny´s rendition has carefully retained the species differentiation marked in the Greek original, although it seems that Pliny does not perceive too many variations in the various horsetails described.(100)

Natural History, XXVII.142 (**Herba tragos quam aliqui scorpion vocant**)(101) as compared with Dioscorides, IV.51.1,(102) is of interest beyond Pliny´s employment of Sextius Niger.   The species is **Ephedra distachya** L.,(103), a joint-pine shrub employed (the seeds being pounded in wine) for diarrhoea and excessive menstruation (Pliny´s **sanguinem excreantibus** is not in Sextius Niger/Dioscorides).   Ephedrine, the active principle derived from **Ephedra** spp. (especially **E. distachya** and Far Eastern species) remains in use as a treatment for the relief of asthma and hay fever due to its properties as a vaso-constrictor and sympathomimetic.   Ephedras have about 0.5 - 2 percent alkaloids and (depending on species) ephedrine and its isomers are about 30 to 90 percent.(104)

There are instances in Pliny´s drug lore in which parallel sources seem to be piled one on another, so that definite assigment of a single source becomes impossible.   For example, **Natural History,** XXIV.117-120 can be compared with Dioscorides, IV.37, and a presumed common source is Sextius Niger.   The plant described is the blackberry (Pliny: **rubus;** Dioscorides: **batos**) **Rubus fruticosus** L. agg., used, according to Dioscorides and his source, as a hair dye, styptic and dessicant, as well as an anti-diarrhoeal agent, a drug to treat excessive menstruation, an antidote against the bite of poisonous animals, a general wound styptic, a treatment for haemorrhoids and as a general mouthwash.   One can, however, detect traces of several other sources, especially in Pliny: Theophrastos (description of **batos**);(105) Zopyrus (**batos** as a standard astringent);(106) Celsus (various uses of **rubus**);(107) Nicander (flower of the **batos** useful against snakebite).(108)   Another series of a mixed sequence of possible citations by Pliny can be seen by comparing **Natural History,** XXXII.33 with Dioscorides, II.79.2 and then noting

the parallels in Galen (from Apollodorus),(109)
Nicander **Theriaca** and **Alexipharmaca** (from Apollodorus)(110) and wondering which collection of
medical opinions would have given Pliny **both**
Nicander and Apollodorus, or perhaps supposing that
Pliny had texts of the lost works on toxic animals
and plants by Apollodorus before him, as well as
texts of Nicander´s **Theriaca** and **Alexipharmaca.**
　Pliny´s assessment of particular substances in
his drug books of the **Natural History** closely
follows his employment and understanding of his
written sources in both Greek and Latin. In the
case of his use of the Greek texts of Sextius Niger
(who has, in his turn, compiled data probably based
on Solon of Smyrna), Pliny shows a usual care with
the particulars and an occasional lapse that may be
due to his rapid methods of reading and compiling.
More often than not, however, Pliny simply follows
his sources verbatim, with slight variations as he
knows them from other corollary sources, but it
seems clear that he depended very heavily indeed on
Sextius Niger for his drug lore. There is some indication also that Pliny may have had a text of
Solon of Smyrna and the compilations by Xenocrates,
but the **Natural History** rarely reflects the employment of pharmacological texts that predate Augustus.
Exceptions may be the Greek texts of Theophrastos
and Nicander, and perhaps Apollodorus, but Pliny´s
drug lore from his Greek sources is generally derivative from a long tradition of pharmacological
doxography, represented by both Sextius Niger and
Solon of Smyrna. Pliny is not the best of medical
botanists, although he does often mirror a common
knowledge of herbal plants that would be well known
in Italy: his truncation of Sextius Niger´s accounts
of non-Italian plants continually suggests that
Pliny did not understand the often subtle distinctions necessary to comprehend different species as
they grew in the eastern Mediterranean. And as
mentioned, there are signs of scribal meddling in
the history of the Latin texts of the **Natural History,** which will further obscure those texts which
cannot be checked against extant Greek sources.

NOTES

Abbreviations and short titles
Celsus: F Marx, ed., **A.Cornelius** Celsi quae
supersunt, Leipzig, Teubner, 1915 (**CML I**)
CMG: Corpus Medicorum Graecorum
CML: Corpus Medicorum Latinorum

Diocles: M Wellmann, ed., **Die Fragmente der sikelischen Ärzte Akron, Philistion und des Diocles von Karystos,** Berlin, Weidmann, 1901 (Fragmentsammlung der griechischen Ärzte, I)
Dioscorides: Max Wellmann, ed., **Pedanii Dioscuridis Anazarbei De materia medica,** Berlin, Weidmann, 1906-1914; 3 vols., reprinted 1958
K.: C G Kühn, ed., **Claudii Galeni Opera Omnia,** Leipzig, Cnoblochius, 1821-33; 20 vols. in 22, reprinted Hildesheim, G Olms, 1964-65
Littré: E Littré, ed., and trans. **Oeuvres complète d´Hippocrate,** Paris, 1839-1861; 10 vols., reprinted Amsterdam, Hakkert, 1973-80
NH or Pliny, **NH:** various eds. of the volumes of the Budé edition (Paris: Société d´Edition ´les Belles Lettres´), viz.:

Jean Beaujeu, ed., **Pline l´Ancien Histoire Naturelle,** Livre I (1950)

Henri le Bonniec and Andre le Boeuffle, eds., **Pline l´Ancien Histoire Naturelle,** Livre XVIII (1972)

J André, ed., **Pline l´Ancien Histoire Naturelle,** Livre XIX (1964)

J André, ed., **Pline l´Ancien Histoire Naturelle,** Livre XX (1965)

Jacques André, ed., **Pline l´Ancien Histoire Naturelle,** Livre XXI (1969)

Jacques André, ed., **Pline l´Ancien Histoire Naturelle,** Livre XXII (1970)

Jacques André, ed., **Pline l´Ancien Histoire Naturelle,** Livre XXIII (1971)

Jacques André, ed., **Pline l´Ancien Histoire Naturelle,** LivreXXIV (1972)

Jacques André, ed., **Pline l´Ancien Histoire Naturelle,** Livre XXV (1974)

A Ernout and R Pépin, eds., **Pline l´Ancien Histoire Naturelle,** Livre XXVI (1957)

A Ernout, ed., **Pline l´Ancien Histoire Naturelle,** Livre XXVII (1959)

A Ernout, ed., **Pline l´Ancien Histoire Naturelle,**Livre XXVIII (1962)

A Ernout, ed., **Pline l´Ancien Histoire Naturelle,** Livre XXIX (1962)

Alfred Ernout, ed., **Pline l´Ancien Histoire Naturelle,** Livre XXX (1963)

Guy Serbat, ed., **Pline l´Ancien Histoire Naturelle,** Livre XXXI (1972)

E de Saint-Denis, ed., **Pline l´Ancien Histoire Naturelle,** Livre XXXII (1966)
Oribasius: J Raeder, ed., **Oribasii Collectionum Medicarum Reliquiae,** Leipzig, Teubner, 1928-1933; 4

vols., reprinted Amsterdam, Hakkert, 1964 (**CMG** VI.1, 1-2, 2)
Paul of Aegina: I L Heiberg, ed., **Paulus Aegineta**, Leipzig, Teubner, 1921-1924; 2 vols. (**CMG** IX 1-2)
PH: Pharmacy in History
RE: Real-Encyclopädie der classischen Altertums-wissenschaft

1.   Pliny, **NH**, XXIX.14 (quoting Cato on Greek physicians). Alan E Austin, **Cato the Censor**, Oxford, 1978, pp. 170-171.   "(Cato)...preferred to rely on traditional and familiar recipes, taking from the Greeks only a few herbal remedies similar in kind" (p.171). E.g., **NH**, XVIII.205
2.   John Scarborough, **Roman Medicine**, London, 1969, pp. 63-65
3.   E.g. **NH**, XXVII.131, as examined in J H Phillips, "Juxtaposed medical traditions", **Classical Philology** 76, 1981, pp.130-132. Cf. Celsus, V.26.35C and VI.7.1D
4.   J Stannard, "Medicinal plants and folk remedies in Pliny, Historia Naturalis", **History and Philosophy of the Life Sciences** 4, 1982, pp. 3-23
5.   Caelius Aurelianus, **Acute Diseases**, III.16.134 (ed. Drabkin, p. 386) says that Niger was a friend of Julius Bassus (n.6 below), which probably dates him to about 10-40 AD. Galen thinks well of him, as do Pliny and Dioscorides (independently).   John   Scarborough and Vivian Nutton, "The Preface of Dioscorides´ Materia Medica: introduction, translation, commentary", **Transactions and Studies of the College of Physicians of Philadelphia**, n.s. 4, 1982, pp. 187-227: 206, and Max Wellmann, "Sextius Niger. Eine Quellenuntersuchungen zu Dioskorides", **Hermes**, 24, 1889, pp.530-569
6.   Julius Bassus **fl.** c. 10-40 AD. Scarborough and Nutton, "Preface", (n.5 above), p. 205
7.   G E R Lloyd, **Science, Folklore and Ideology**, Cambridge, 1983, pp.135-149
8.   W H S Jones, "Ancient documents and contemporary life", in: E A Underwood, ed., **Science, Medicine and History: Essays...in honour of Charles Singer**, London, 1953, 2 vols.; vol. 1, pp. 100-110: 106-108
9.   Esp. eye diseases, skin ailments and rashes, various fevers, and the bites and stings of poisonous animals and insects.
10.   M Grieve, **A Modern Herbal**, New York, 1931; 2 vols. Reprinted 1971. Hans Flück,

Medicinal Plants, trans. (from the German) J M Rowson, London, 1976. Paul Schauenberg and Ferdinand Paris, **Guide des plantes médicinales**, 3rd ed., Neuchâtel and Paris, 1977. Cf. Juliette de Baïracly Levy, **Herbal Handbook for Farm and Stable**, rev. ed., London, 1973

11. The range of modern approaches is merely suggested in the following volumes: G E Trease and W C Evans, **Pharmacognosy**, 11th ed., London, 1978; W H Lewis and M P F Elvin-Lewis, **Medical Botany**, New York, 1977; V E Tyler, L R Brady and J E Robbers, **Pharmacognosy**, 8th ed., Philadelphia, 1981; A F Hill, **Economic Botany**, 2nd ed., New York, 1952.

12. A Sofowora, **Medicinal Plants and Traditional Medicine in Africa**, New York, 1982, and T Swain, **Plants in the Development of Modern Medicine**, Cambridge, Mass., 1972 show the modern counterparts of this venerable tradition throughout the world. One may also cite the rich pharmacy and medical plant lore of pre-Columbian America suggested by Ralph L Roys, **The Ethno-Botany of the Maya**, New Orleans, 1931, reprinted Philadelphia, 1976

13. Jones, "Ancient documents" (n.8 above), p.109, puts it, "In Pliny pharmacology has swamped everyhing else".

14. Welmann, "Sextius Niger" (n.5 above) passim

15. **NH**, XX:1606. XXI:730. XXII:406. XXIII:1418. XXIV:1116. XXV:1292. XXVI:1019. XXVII:602. XXVIII:1682. XXIX:621. XXX:854. XXXI:924. XXXII:990. XXXIII:288. XXXIV:257 (**medicinae**)

16. Esp. Celsus, V.1-17: simples classed by pharmaceutical/physical properties, e.g. V.2: **glutinant vulnus** (19 substances)

17. **Vid.** e.g. the Hippocratic **Wounds**, 11-12 (ed. Littré, VI.410-417)

18. John Scarborough, "Theophrastus on herbals and herbal remedies", **Journal of the History of Biology** 11, 1978, pp. 353-385

19. John Scarborough, "Theoretical assumptions in Hippocratic pharmacology", in: F Lasserre and P Mudry, eds., **Actes de Colloque hippocratique Lausanne 1981**, Lausanne, 1983, pp.275-293: 276-280

20. Max Wellmann, "Das älteste Kräuterbuch der Griechen", in: **Festgabe für Franz Susemihl**, Leipzig, 1898, pp.1-31; and John Scarborough, "Nicander´s toxicology,I: snakes", **PH**, 19, 1977, pp.3-23: 3-4; and "Nicander´s toxicology, II:

spiders, scorpions, insects and myriapods", **PH**, 21, 1979, pp.3-34 and 73-9:4

21. Scarborough, "Nicander I" and "Nicander II" (n.20 above) passim

22. John M Riddle, **Dioscorides**, Austin, Texas, in press (1984) develops this thesis with precision. **Vid.** also Scarborough and Nutton, "Preface" (n.5 above) pp. 190-1

23. Felix Rinne, "Das Receptbuch des Scribonius Largus", **Koberts Historische Studien aus dem Pharmakologischen Institut der Universität Dorpat**, 5, 1896,1-99: 32, reprinted in **Historische Studien zur Pharmacologie der Griechen, Römer und Araber, Leipzig DDR 1968**

24. The basic modern edition is Georg Helmreich, ed., **Scribonii Largi Conpositiones**, Leipzig, Teubner, 1887, but Sergio Sconnochia, **Per una nuova edizione di Scribonio Largo**, Brescia, 1981, has announced preparation of a new text based upon fresh readings obtained from MS Toletanus 98,12.

25. Analysis of Galen's "drug books" shows that he remained uncertain regarding the best method of pharmaceutical classification. The three major tracts by Galen on drugs are **Mixtures and Properties of Simples** in eleven books (XI.459-892, K. and XII.1-377. K.) **Compound Drugs arranged by Location of Ailment**, in ten books (XII.379-1007, K., and XIII.1-361, K.), and **Compound Drugs arranged by Kind**, in seven books (XIII.362-1058). **Mixtures** is a listing of about 440 plants (books I-VIII) and about 250 other substances in the remainder of the tract, and - along with a 'system of degrees' - the organisation principle is an alphabetic arrangement. **Drugs by Ailment** was written down after **Mixtures** (XII.378, K. = **Drugs by Ailment**, I.1) and here Galen provides a crude 'head to toe' ordering, with gout qualifying as the 'bottom' tier. **Drugs by Kind** was written in response to criticism of his earlier drug works (XIII.363, K. = **Kind** I.1) and incorporates the venerated 'treatment by contraries', with numerous earths, plasters, pills, etc., quoted from many authorities, esp. Andromachus.

26. J Stannard, "Pliny and Roman botany", **Isis**, 56, 1965, pp. 420-425

27. E.g., using Sextius Niger: **NH**, XXXII.26 compared with Dioscorides, II.62

28. The only sound Greek text is that ed. by Max Wellmann in vol. 1 of the **Materia Medica**, pp. 1-5. One may **not** trust the "translation" by John Goodyear (1655) in R T Gunther, ed., **The Greek**

Herbal of Dioscorides, Oxford, 1934, reprinted New York, 1959, pp. 1-4. **Vid.** translation (from the Wellmann text) by Scarborough and Nutton (n.5 above), pp.195-7

29. Dioscorides, **Preface, 4. Vid.** commentary by Scarborough and Nutton (n.5 above), pp. 213-217

30. Scarborough and Nutton (n.5 above), pp. 192-4

31. E.g. Philo of Tarsus, **fl.** 20 AD. **Vid.** refs. in Scarborough and Nutton (n.5 above) p. 193, n.24

32. Hippocrates: Dioscorides, III.59 and IV.168. Theophrastos: Dioscorides, III.74 and V.108. Nicander: Dioscorides, III.29 and IV.99. Comparison of Dioscorides, II.169.3 with Galen, **Properties of Foods**, II.63.2 (ed. G Helmreich, **Galeni De Alimentorum Facultatibus**, (Leipzig, 1923; in **CMG** V.4.2) p.326) and the Hippocratic **Diseases**, II.38 (ed. Littré, VII.54) with Pliny **NH** XXII.71 (all on mallow and asphodel) shows a deeply embedded and very ancient source for all four authors. It is significant that Galen, **Properties of Foods**, II.63.1 (ed. Helmreich, p.325) quotes from Hesiod, **Works and Days**, 41. The "old poets" were important doxographical sources in their own right for plant names. **Vid.** also n.36 below. Pliny, **NH** XXII.73, acknowledges this tradition specifically: **Asphodelum ab Hesiodo quidam halimon appellari existimavere, quod falsum arbitror**

33. Dioscorides, Pref.1, trans. Scarborough and Nutton (n.5 above) p.195

34. Dioscorides, Pref.1-2, trans. Scarborough and Nutton (n.5 above) p.195. **Vid.** commentary, ibid.,pp.202-208 for details on Iollas, Heraclides of Tarentum, Crateuas, Andreas, Bassus, Niceratus, Petronius, Niger, Diodotus and Asclepiades.

35. Ibid.,pp.198-9 (commentary)

36. Athenaeus, **Deipnosophistae**, quoting (for example) Aristophanes, Callimachus, Menander and Nicander. **Vid.** index entries in the standard modern edition of the Greek text of the **Deipnosophistae**, ed. Georg Kaibel, Leipzig, Teubner, 1889-1890; 3 vols.

37. Max Wellmann, "Beiträge zur Quellenanalyse des Älteren Plinius", **Hermes**, 59, 1924, pp. 129-156 esp. 129-132.

38. **NH**, XX, XXII, XXIV.124ff. and XXVI.82f.

39. Wellmann, "Plinius" (n.37 above) p. 135, n.7

40. **NH**, XX.85

41.   **NH**, XIV.76, e.g. Hicesius.   Cf. **NH**, XX.35.   The Hicesius citation in **NH**, XXVII.31 (on the cabbage) stems, by contrast, from Xenocrates, and behind Xenocrates is probably the physician Aristagoras, listed **with** Xenocrates in the index of authorities for **NH**, XXIX and XXX.

42.   **NH**, XX .103 and XXIV.186

43.   Max Wellmann, **Die Pneumatische Schule bis auf Archigenes**, Berlin, 1895, p.13

44.   **NH**, XX.17 (Chrysippus)

45.   **NH**, XX.19 (Democritus, Dionysius, Diocles).   Cf. Diocles, Frg. no. 156 (ed. Wellmann, p. 192), which Wellmann suggests comes to Pliny through Julius Bassus, quoting Dionysius.   Cf. also Gargilius Martialis, XXV (ed. V Rose, **Plinii Secundi** quae fertur una cum Gargilii **Martialis Medicina**, Leipzig, Teubner, 1875, p. 173, on turnips, who either borrows directly from Pliny, **NH**, XX.19 or uses the same text (Julius Bassus?) employed by Pliny.

46.   **NH**, XX.28 (Democritus)

47.   **NH**, XX.35 (Hicesius).   Cf. **NH**, XXVII.31

48.   **NH**, XX.85, 113, 119, 185-195 (Erasistratus).   Cf. **NH**, XIX.167 and Dioscorides, III.56

49.   **NH**, XX.187: **laudatissimum**.   XX.190: **utilissimum**

50.   Evenor: **NH**, XX.187, 191.   Iollas:NH, XX.187

51.   Dioscorides, III.56 (on anise).   Galen, **Mixtures and Properties of Simples**, VI.1.4 (XI.833 K.).   **Simples**, VI preface (XI.797 K.) with Niger among many authors named: was Galen using the same pharmaco-doxographical collection?   Symeon Seth, I: on the anise (ed. B Langkavel, **Simeon Sethi Syntagma De Alimentorum**, Leipzig, Teubner, 1868, pp. 23-4.

52.   E.g.   Oribasius, **Medical Collection**, XIV.45.1 (ed. Raeder, II, p.217), quoting from a collection of data by Zopyrus, and XIV.50.2 (ed. Raeder, II, p.223), again from Zopyrus.   Cf. Celsus, II.31.   The "anise" section in Paul of Aegina, VII.3 s.v. is derivative from Galen.

53.   **NH**, XX.185 and 186 (Pythagoras). XX.191-194 (Tlepolemus).   XX.193 (Heraclides).

54.   **NH**, XX.189 compared with Dioscorides, III.56

55.   Wellmann, "Plinius" (n.37 above). p.142

56.   Galen (quoting Andromachus), **Compound Drugs arranged by Location of Ailment**, III.1 (XII.630 K.) and **Compound Drugs arranged by Kind**, VI.13 (XIII.928 K., viz. ed. Silon).

57.   **NH,**   XX.221.   Cf. **sunt   qui**   XX.242 (Castor) and   231 (Niger).   Cf.   Dioscorides, II.119.
58.   **NH,**   XX.220: **miror quare difficulter in Italia nasci tradiderit id Solon Smyrnaeus.**
59.   **NH,**   XX.235: **adiecit   Solo,   ne   quod omittamus.**
60.   Generally **NH,** XXVII-XXXII
61.   Max   Wellmann,   "Xenocrates   aus Aphrodisias", **Hermes,** 42, 1907, pp. 614-629
62.   **Vid.** the confused and confusing melange of   "sources"for   pseudo-Democritus   in   Franz Susemihl, **Geschichte der griechischen Literatur in der Alexandrinerzeit,** Leipzig, 1891-2; 2 vols., vol. I, pp. 483-4.   Cf. the sage remarks by P M Fraser in **Ptolemaic Alexandria,** Oxford, 1972; 3 vols.,   vol.II,   p.641 (n.536) and   pp.   644-645 (n.548).
63.   Esp. interesting for suggesting how the scholia on Nicander can occasionally help in understanding (and correcting) Pliny´s text is I Cazzaniga, "Note critico-filologiche, III: Plinio il Vecchio XXX 35 e la tradizione di Schol. Nicandro Ther. 372", **Studi Classici e Orientale,** 14, 1965, pp. 16-19
64.   Wellmann,   "Plinius" (n.37 above)   pp. 143-156.   Cf. M Geymonat, "Una prefazione in senari al trattato di Valgio Rufo sulle erbe?", **La Parola del Passato,** 29, 1974, 256-261
65.   A Ernout, "La magie chez Pline l´Ancien", **Latomus,** 70, 1964, pp. 190-195
66.   **Vid.** John Scarborough, "The drug lore of Asclepiades of Bithynia", **PH,** 17, 1975, pp. 43-57
67.   Now firmly dated back to the second century BC.   Elizabeth Rawson, "The life and death of Asclepiades of Bithynia", **Classical Quarterly,** n.s. 32, 1982, pp. 358-370
68.   Cajus Fabricius, **Galens Exzerpte aus alteren Pharmakologen,** Berlin, 1972, pp. 192-198
69.   J G Sprengel, **De Ratione, quae in Historia Plantarum inter Plinium et Theophrastum intercedit,** Marburg, 1890
70.   L Renjes, **De Ratione, quae inter Plinium Naturalis Historia lib. XVI et Theophrastum libros de Plantis intercedit,** Rostock, 1893
71.   Wellmann, **Pneumatische   Schule** (n.43 above) p. 26, n.3 (cont. from p. 25)
72.   H Stadler, **Die Quellen des Plinius im 19 Buch der Naturalis Historia,** Munich, 1891; Simon Sepp, **Pyrrhoneische Studien,** Freising, 1893, esp. pp. 53-58

73.   Fabricius, **Galens Exzerpte** (n.68 above)
p. 226, following F Kudlien, "Xenocrates, Nr.8", **RE**,
2nd ser., IX part 2, Stuttgart, 1967, cols.
1529-1531
74.   Manfred Ulllmann, "Xenocrates Nr. 7",
RE,, Supplementband XIV, Stuttgart, 1974, cols.
974-977: 975
75.   Galen, **Mixtures and Properties of
Simples**, X.1 (XII.248, K.)
76.   Oribasius, II.58 (ed. Raeder, I.)
77.   Galen, **Mixtures and Properties of Simples**
X.2.4 (XII.261, K.)
78.   Ullmann (n.74 above) col.976
79.   John Scarborough, "Diphilus of Siphnos
and Hellenistic medical dietetics", **J. Hist. Med.**,
25, 1970, pp. 194-201
80.   Dioscorides, **Preface**, 5: "For I have
exercised the greatest precision in getting to know
most of my subject through direct observation, and
in checking what was universally accepted in the
written records and in making enquiries of the
natives in each botanical region." trans.
Scarborough and Nutton, (n.5 above), p. 196
81.   Dioscorides, **Preface**, 2. **Vid.** also n.
34 above
82.   Dioscorides, **Preface**, 2, trans. Scarborough and Nutton (n.5 above) p. 196
83.   Dioscorides, **Preface**, 2, trans. Scarborough and Nutton (n.5 above) p. 196
84.   For example, **NH** XXVII.4-142 (**aconitum** to
tragopodon)
85.   Scarborough and Nutton (n.5 above), pp.
210-212 (on aloe). **Vid.** also John Scarborough,
"Roman pharmacy and the eastern drug trade: some
problems as illustrated by aloe", **PH**, 24, 1982, pp.
135-143
86.   Scarborough and Nutton (n.5 above) p. 209
(commentary on spurge-resin). **Vid.** Dioscorides,
III.82
87.   Scarborough and Nutton (n.5 above) pp.
209-210 (commentary on the two kinds of St John´s
wort).
88.   Scarborough and Nutton (n.5 above) pp.
206-208 (commentary on Asclepiades and the **anarmoi
onkoi**).
89.   ed. A Ernout, **NH**, XXVII.76 (pp.45-46).
App. crit. p.45 suggests **gustu languido** for the
lacuna (from Dioscorides)
90.   ed. M Wellmann,II, p. 184
91.   ed. A Ernout, **NH**, XXVII, p. 94: "Plante
non identifée", following all dictionaries of Greek

and Latin.

92.   Nicander, **Theriaca**, 647.   Diocles, Frg. 149 (ed. Wellmann, p. 191). Pliny, **NH**, XXIII.131

93.   Paul of Aegina, VII.3 s.v. (ed. Heiberg II, p. 211)

94.   ed. Jacques André, **NH**, XXII.51-2 (pp.39-40)

95.   ed. Wellmann, II, pp. 188-9

96.   ed. André, **NH**, XXII, p. 93 (commentary)

97.   Ibid., p. 94 (accepted with reservations).

98.   Among a great number, e.g. Dioscorides, IV.86 and Pliny, **NH**, XXVII.23; Dioscorides, IV.88 and **NH**, XXV.160; Dioscorides, IV.89 and **NH**, XXV.161; Dioscorides, IV.90 and **NH**, XXV.162; Dioscorides, IV.91 and **NH**, XXV.159

99.   This plant turns up in the papyrus Vindobonensis D. 6257 as the Demotic **ertyn**; the papyrus emerges from Roman Egypt, but records data from far earlier eras. Pap. Vinob. D. 6257 Bk. A, col. x+II, 24

100.   **NH**, XXVI.134: **faciunt et aliam hippurim.**

101.   Cf. **NH** XIII.116

102.   Cf. **NH**, XXI.119

103.   ed. Ernout, **NH**, XXVII, p. 115 (commentary)

104.   Trease and Evans (n.11 above) pp. 565-7

105.   Theophrastos, **Historia Plantarum**, III. 18.4

106.   Oribasius, **Medical Collection**, XIV.61 (ed. Raeder, II, p. 231, from Zopyrus´ **Astringents**).

107.   Celsus, II.33.4 and III.19.2 (pounded leaves as powder + wine as repressant chest plaster); IV.23.2 (**rubus** + plantain as anti-diarrhoea medicine); and IV.26.8 (**rubus** decoction for anti-diarrhoea treatment: part of a section of drugs from orchard fruits).

108.   Nicander, **Theriaca**, 839

109.   Galen, **Antidotes**, II.14 (XIV.182, K.)

110.   Nicander, **Theriaca**, 700-713, and **Alexipharmaca, 585.**

Chapter Six

PLINY ON PLANTS: HIS PLACE IN THE HISTORY OF BOTANY

A G Morton

In one of the rhetorical passages which Pliny liked
to scatter at intervals among the "twenty thousand
facts" he claimed to have included in the **Natural
History** he becomes enthusiastic about the majesty of
the **Pax Romana,** which fostered wide movements of
useful plants about the world, bringing liquorice
from Russia, the water-dock from Britain, the spiny
euphorbia from North Africa and a species of sage
(**Salvia aethiopis**) from Ethiopia (**Natural History,**
27.1). It was their utility to mankind that
excited Pliny's interest in plants, not general
principles that arise from their scientific study.

Although the **Pax Romana** was not in reality noted
for its peacefulness, the social and administrative
unity which it encompassed still preserved in
Pliny's time an unbroken connection with classical
and earlier Greek civilisation; its records of
philosophy, science and medicine were accessible to
Pliny and could be used by him in compiling the
Natural History. It is this direct link with
ancient Greek science which gives his work special
interest for us.

Pliny did not claim originality of discovery or
thought, but he prided himself on using the foremost
authorities to give a popular well founded account
of his subject matter. To expect original
observation from him would therefore be unfair, but
we are justified in examining how accurately and
effectively he transmitted the knowledge of his
time. It is primarily from this point of view that
I am going to look at Pliny's account of plants and
the influence he had on the subsequent development
of the science of botany.

Before proceeding to the main theme, I should
like to place it in context by a short digression on
botany before Pliny. As he remarked, with plants

in mind, "It is impossible sufficiently to admire
the pains and care of the ancients, who explored
everything and left nothing untried" (**Natural
History**, XXIII.112). The development of a definite
science of botany from such traditional empirical
knowledge took place (in the West) at about the same
time as the other natural sciences were in process
of formation, between about 500 BC and 200 BC in
Ionia and Greece.

Evidence of an approach to a science of plants
first occurs in writings in the Hippocratic corpus
derived largely from the island of Cos. Lists of
medicinal and food plants (the latter were in fact
regarded primarily as medicinal also) were compiled,
sometimes with descriptive phrases that point to
their identity; these lists constitute the earliest
known body of herbal knowledge in the West. At the
same time attempts were made to compare the
structure and physiology of plants with that of
animals and man. Agriculture expanded during this
period and led to the production of a considerable
literature on various aspects of farming, written by
the more intelligent and enterprising land-owners to
help their fellows, and containing many observations
on the behaviour of plants. The expansion of trade
and the movements of military conquest brought
acquaintance with increasing numbers of exotic
plants and plant products, and some of these foreign
plants were brought into cultivation.

It was to botanical information from these three
sources, and from his observations and research,
that Theophrastos applied the analytical-logical
methods of Aristotelian philosophy to create, in his
lectures on plants at the Lyceum, the first synoptic
treatment of scientific botany. We are fortunate
in possessing a substantially complete record of his
botanical teaching, written down about 300 BC when
he was seventy. The two books, usually called by
titles given to them by a later editor, **Enquiry into
Plants** and **Causes of Plants**, deal respectively with
what today might be called descriptive botany and
physiological botany. I shall refer to their
contents more specifically later on in relation to
Pliny.

A twentieth-century botanist reading Theophrastos
is struck with astonishment and admiration: not only
by the wealth of detailed and accurate observation
of plants and their behaviour, anticipating at least
in embryo almost every branch of modern botanical
enquiry, but even more by the deep critical
discussion of what were, and in developing form have

87

remained, fundamental questions of botanical theory.
  The writings of Theophrastos were the first
flowering of ancient botany as a science, but they
were also its culmination. A decline of interest
in theoretical science, which historians have
recognised as a general tendency in the Hellenistic
and Roman world, caused a change in attitude to
Theophrastos. The impact of his theoretical
synthesis in botany is scarcely mentioned by any
later writers. His great authority continued
however to influence a succession of Greek and Roman
herbalists and pharmacologists, who based themselves
on his methods of description and differentiation of
plants, although usually falling far short of his
practice at its best. In agriculture,
Theophrastos' systematic treatment of plant
reproduction and of the effects of the weather, soil
and cultivation on plant growth, was embodied in
standard techniques and practices, passed on by the
writers of agricultural handbooks in empirical form,
but not explicitly linked to his leading ideas.
  This sketch will I hope be sufficient to indicate
the botanical background to Pliny's work on plants,
written more than 300 years after the death of
Theophrastos, when his genius was acknowledged, but
was already thought of as that of a
philosopher-theorist, somewhat remote from the
interests of practical men.
  It seems surprising that Pliny devoted sixteen
books, that is, about two fifths, of the **Natural
History** to plants, since there is no evidence that
his omnivorous curiosity about the world was more
engaged by the plant kingdom than by many other
topics. Possibly the simple explanation is that
the technical literature relating to plants was more
copious that in other areas because of the
predominance of plants in supplying the material
needs of society. Metals might change the
foundations of the social order, but plants kept it
in being. A very extensive literature on plants
was at his command, if we are to accept at face
value the lists of his authorities, of which he is
so naively proud. The names of almost all the
botanists, herbalists, agricultural writers, medical
writers, mentioned anywhere in ancient sources, seem
to find a place. The works of these authors (many
now lost) were presumably accessible to him, but
whether he actually consulted them all is not easy
to decide. We know that the greater part of his
work was compiled from a few major authorities - as
I shall discuss below - but his occasional citation

by name and the considerable amount of factual
material he includes for which no source can be
identified, suggest that he did not simply take over
references from other people.

A very interesting example of what seems
genuinely new matter is Pliny´s account (**Natural
History**, XVII.42 et seq.) of the agricultural use of
marl in Gaul and Britain, where the Celtic words for
the different types of marl (**marga, acaunamarga,
glisomarga**) are quoted, perhaps from personal
knowledge gained whilst travelling on military
business, but more probably derived from the works
of Pompeius Trogus, who was a native of Gaul, and
although from the south would be familiar with
farming practices in the north .

It is quite clear thay Pliny recognised
Theophrastos as the supreme authority on plants.
He indicates this in his lists of authorities for
the books on plants, by placing the name of
Theophrastos first among foreign (non-Roman)
authorities in ten cases, and second after either
Hesiod, Herodotus or Democritus in the other six
cases (probably because the first named were ancient
authors by comparison). In three cases the name of
Aristotle follows that of Theophrastos, which agrees
with the tradition that Aristotle also wrote on
plants. Pliny also mentions Theophrastos by name
more than a dozen times in the books on plants on
the **Natural History**, always with the greatest
respect, calling him "one of the most celebrated
Greek authors" and dating his works to 314 BC (which
is not far out). Pliny cannot resist quoting from
Theophrastos a rhizotomist´s fable about an unnamed
and undescribed herb which confers the capacity for
repeated sexual congress (up to seventy times!), but
he adds the extenuating phrase that Theophrastos is
"in general a weighty authority". Manuscripts of
Theophrastos were evidently to be seen without
difficulty, for Pliny speaks of consulting more than
one copy.

The strongest evidence for Pliny´s reliance on
Theophrastos comes however from the content of the
books on plants (**NH**, XII-XXVII) in the **Natural
History**. The first eight books (**NH**, XX-XIX)
describe plants in general and represent Pliny´s
treatment of botany. The material throughout is
heavily dependent on Theophrastos (especially on the
Enquiry), and many passages are directly transcribed
from him. Indeed it is clear that Pliny intended
to follow the general arrangement of the **Enquiry**,
but the logical thread of Theophrastos becomes

repeatedly lost in Pliny´s compulsive diversions and anecdotal embellishments. Nevertheless, the literal connection with Theophrastos is abundantly evident, and has been assiduously documented by learned commentators over the years. It is also clear that Pliny had read the **Causes of Plants** but makes less direct use of it, perhaps because the subject matter is less capable of expresssion as a series of facts, for which Pliny had a collector´s passion.

The remaining eight books on plants (**NH,** XX-XXVII) were devoted by Pliny to medicinal plants, a deliberate division of the subject matter, he says, for the convenience of two classes of reader having different requirements. The strict botanical interest of these books is narrower, but Pliny manages to include a good deal more in the way of botany than mere descriptions of plants. For the material in these books, Pliny (like his contemporary Dioscorides, writing unknown to him) drew very largely on the **Treatise on Greek Medicine,** an alphabetical list of drug plants (in Greek) compiled by Sextius Niger, who himself drew on the main Greek herbal-pharmacological tradition stemming from Theophrastos, Diocles of Carystos, Apollodorus of Alexandria, Andreas of Alexandria and Crateuas, all of whom Pliny included in his list of authorities.

In these books there is naturally information from a wider range of sources, but Pliny cites Theophrastos in many passages, and very significantly finishes the last book with a long excerpt from the conclusion of the final book of Theophrastos´ **Enquiry** - as if to hammer home that Theophrastos is his model.

Although Pliny´s foundation is Theophrastos, his use, in his own diverting fashion, of so many other sources, written or verbal, gives his work perennial interest. There is hardly a page in the books on plants which does not give a modern botanist some insight into ancient plant knowledge or technology, or rouse some curious problem concerning the source or accuracy of his facts, or the identity of his plants. He is eminently an author for browsing in, but I must resist the temptation to explore his fascinating bye-ways in order to look at the general character of his survey of ancient botany and its place in the history of science.

First, a word on the question of whether he made any original contributions to botany. In an autobiographical insertion (**NH,** XXV.9), he says that he personally looked at all but a few plants in the garden of Antonius Castor in Rome, the supreme botanical expert of the time. This would be just the kind of opportunity that a busy man like Pliny would seize to learn something about plants, and no doubt he found other occasions to have plants shown to him by experts or even country herb-gatherers. Elsewhere (**NH**, XXVI.11) he speaks of the degeneration of medicine, now medical men preferred to learn about herbs from lectures instead of learning from experience, the best teacher, and seeking plants in the wild at the proper season. There is however nothing to suggest that Pliny ever had time to look at wild plants for himself. His descriptions, even of very common plants, are invariably copied. Where he describes the same plant as Theophrastos, his description is nearly always less complete and less precise. Even the common coltsfoot (**Tussilago farfara**) he describes as without flower, although in fact it flowers and seeds early in spring before the leaves appear, as was known to Dioscorides, who corrects what he says is a common error. When Pliny gives his own, almost lyrical description of the beauty of the gromwell (**Lithospermum officinale**), he reveals that he only saw a picked specimen, brought to him, not one growing in the ground.

In two cases, however, Pliny certainly contributed new botanical facts acquired during military service. His reference (**NH**, XVII.30.121) to beans growing on islands in the North Sea must be the first mention in literature of the exclusively northern maritime pea (**Pisum maritimum**).(1) In Spain, Pliny distinguished two plants used for making ropes which bore the name **spartum** (**NH**, XIX.26; XXIV.65). One with "yellow flowers and seeds in pods" is correctly identified by him as **genista**, the plant known to the Greeks as **sparton** (broom, **Spartum junceum**), and used from earliest times for ropes. The other **spartum**, cultivated by the Carthaginians in Spain, is said by Pliny to be a rush-like plant (**iuncus**) of arid ground, and is clearly the esparto grass (**Stipa tenecissima**), of which he describes the cultivation, technology and economic importance. Here we have Pliny at his shrewdest and best, relating, as he occasionally does in other places, some fact that he has seen or learned for himself. That he ever made real botanical investigations seems, however, most unlikely.

Theophrastos on the other hand was an original

observer and scientist who brought the judgement of experience to every aspect of botanical investigation. Pliny had no such basis of experience, and this, even more than the differences in intention, temperament and intellect, explains his blindness to the great theoretical advances which entitle Theophrastos to be called the father of scientific botany. The vital contributions made by Theophrastos to the methods and general principles of botany were either ignored by Pliny or were misunderstood and robbed of their meaning and importance. A full discussion of these questions would be too lengthy and is not necessary for my purpose. I shall therefore refer only to a few major points by way of illustration.

Theophrastos begins by defining botany, in surprisingly modern terms, as the investigation of the form, reproduction and behaviour of plants. His subsequent treatment conforms to this scheme and is based on the scientific-philosophical methodology of Aristotle (in practice close to that of modern science). Theophrastos departs very significantly from Aristotle, however, by almost entirely excluding teleological interpretations and viewing plant life and development in relation to material causes alone. All this is quite foreign to Pliny, who adopts Stoic doctrine in its most superficial form, and treats plants as mysterious and wonderful products of nature, designed and adapted for the prime purpose of satisfying the needs of human beings. Thus in a simple list of mountain plants. Pliny includes "broom plants produced by nature to dye cloth". Elsewhere he reveals his shallow conventional philosophy in a passing reference to Epicurus as "that master of leisured ease" (**otii magister: NH**, XIX.51).

Theophrastos opens his **Enquiry into Plants** with a carefully argued analysis of plant form. Individual tissues and organs are defined by function, position and interrelations, and Theophrastos has a special terminology for the various parts based on analogical parts in animals. He makes clear that plants are **sui generis** and have only analogy to like-named parts of animals. This profound discussion is the historical beginning of plant morphology, but is not represented in Pliny, except perhaps by the confusing statement that "in general the bodies of trees, as of other living things, have in them skin, flesh, sinews, veins, bones and marrow" (**NH**, XVI.181).

It is interesting that Pliny refers once (**NH**,

XVII.153) to the Hippocratic idea, repeated by
Aristotle, that the vital soul of plants resides in
the pith (medulla), a view which Theophrastos
probably rejected - he does not mention it - since
he knew that pith is often absent in healthy plants.
   The methods of identifying and classifying plants
were deeply considered by Theophrastos, who proposed
numerous morphological characters (**diaphorai,
differentia**) obviously based on discriminating
examination of a wide range of plants and their
parts, which he suggested would be useful in
classification. Almost all his ideas have since
proved valuable and become part of the fabric of
botany. In addition Theophrastos recognised a
considerable number of natural groupings of plants,
to some of which he gave names expressing their
relationships to particular species, which he
regarded as type (for this grouping). This is
another area in which Pliny had no interest and
where he conveys nothing of the important concepts
evolved by Theophrastos, although he reproduces some
of the categories incidentally. Pliny´s own rare
groupings of plants are arbitrary, even bizarre, and
when he follows the separation, made by
Theophrastos, of cereals and legumes as natural
groups, he is confused and omits the differing
character of their seeds, emphasised by Theophrastos
(**NH**, XVIII.49 et seq.).
   The nature and function of the flower were
difficult questions for the early botanists.
Although Theophrastos recognised the general
association between flower and fruit or seed in most
plants (but not in all, which was what puzzled him),
the existence of sex in plants was problemetical
(and remained so for two thousand years after his
death). Fertilization of the female date-palm by
dust from the male was very well known to
Theophrastos. The technique of artifical
fertilization was described by him in both the
Enquiry and the **Causes**, and the similarity to sexual
union was noted: but it remained a special case,
which could not be harmonised with the situation in
plants in general, as Theophrastos explains in a
very interesting passage (in **CP**, III.18.1) which
shows the depth of his thought. When Pliny came to
the date palm he repeats Theophrastos literally but
cannot resist inserting an imaginative and somewhat
indelicate description of sexual encounters between
male and female trees in the palm grove. He does
not, however, appear to see the need for any
unifying hypothesis, although he quotes dogmatically

Aristotle´s view that all plants contain both sexes united in them.

In keeping with his philosophic outlook, Pliny regarded flowers as the material for ceremonial wreaths or medical prescriptions and not, as Theophrastos did, as specific structures of a certain class of plants, displaying a range of form (differentia) by which they can be distinguished and classified. It is therefore strange to find that in three flowers, lily, rose and saffron-crocus, where Pliny copies the descripion given by Theophrastos, he adds details of floral structure not given by the latter (**NH**, XXI.14-15, 22, 33). Pliny´s account of the lily contains an unmistakable reference to the pistil and stamens, called by him **pilum** (pestle) and **stamina** (from **stamen**, the upright warp of the ancient loom). Because the **Natural History** was widely known during the middle ages and renaissance, the word ´stamen´ began to be used during the 15th and 16th ceturies as a botanical term for the organ that was becoming recognised as regularly present in flowers. The word was finally fixed in botanical nomenclature by Linnaeus in the 18th century. For some odd reason Pliny´s description of the lily was very popular with preachers and encyclopedists during the middle ages and was often quoted without acknowledgment - perhaps the attraction was the description of the shining white petals, which in fact was entirely cribbed from Theophrastos. Pliny cannot be credited with inventing the technical term for stamen, since he used a different word, **apices**, to describe the stamen of the rose. He does not mention the pistil in the rose, whilst in the saffron-crocus he refers only to the stigma, but in this case terms it **capillus** (hair), not **pilum**. So Pliny was certainly not recognising and naming new morphological categories, as Theophrastos undoubtedly did. Was he recording original observations? We do not know. It is possible, but, alas, Pliny´s literary habits make it more probable that he is quoting from some source unknown.

To conclude this part of my discussion I shall refer very briefly to Pliny´s treatment of agricultural and garden plants, their cultivation and relation to weather conditions, type of soil, and the environment in general; these matters are treated mainly in books XVII, XVIII and XIX. The influence of Theophrastos is present throughout, but Pliny makes less direct use of him than elsewhere, and does not expound the remarkable theoretical ideas in relation to plant nutrition and metabolism and to the adaptation of plants to the environment, concepts developed particularly in the **Causes of Plants**, which clearly Pliny did not fully understand. Instead, Pliny relies more on later agricultural writers: Androtion, Cato, Varro, Virgil´s **Georgics**, Mago the Phoenician (whose works had been translated into Latin), and above all on Columella. The agricultural parts of Pliny continued to exercise some practical guidance till late in the middle ages, although they are much inferior to Columella in this respect.

My insistence on Pliny´s failure to transmit the real scientific content of Theophrastos may seem ungracious and unreasonable. But you will of course understand that I am not reproaching Pliny for doing what was beyond his powers, but am commenting on an accident of history. Pliny could not have dreamt that less than two centuries after his death the text of Theophrastos would be effectively lost, and that the books on plants in his **Natural History** would perforce become, what he never intended them to be, the only record of the first great scientific treatise on botany. For almost 1400 years learned men knew only, mainly on Pliny´s authority, that Theophrastus wrote about plants, and ancient botany was represented by Pliny on plants and by the herbs in the **Materia Medica** of Dioscorides.

By miraculous good fortune manuscripts of Theophrastos´ two works on plants were discovered in the 15th century, just in time for their remarkable scientific content to be assimilated into the reviving study of botany and to become part of the evolution of modern science. Theodore Gaza, the first translator of Theophrastus into Latin, in about 1546 (printed 1483) used Pliny to elucidate some of the difficulties of the text.

How then are we to estimate Pliny´s influence on botany during nearly a thousand years when botany was at a standstill or declining, and during the slow revival in the four centuries before the renaissance? I think that with all his weaknesses and omissions Pliny did one very important thing: he kept alive the conception of botany as a broad, unitary science of plants, not to be reduced to a knowledge of drug-plants or the operations of cultivation, a conception that originated with Aristotle and Theophrastos at the Lyceum. The wide ranging mass of information about plants which he gathered, often uncritically and without full understanding, nonetheless preserved a level of botanical continuity, for which he is entitled to recognition and gratitude.

In this context it is to be noted that the plants mentioned by Pliny represent (as do those in Theophrastos) a botanical collection, a reasonable section across the whole realm of plants; of course, there is some general bias towards useful plants, but not to one particular use. It is doubtful whether Pliny added any new plants to the basic corpus of about 800 plants known to the Greeks, and which he found in Theophrastos and Sextius Niger. Pliny mentions about 1300 plant names, but many are synonyms; his total number of species probably does not exceed 800. Pliny's descriptions of plants were almost all copied from his authorities and many are too vague and defective for identification.

I have not touched, in this discussion, on the vast miscellany of facts, fables and gossip, more or less connected with plants, which makes Pliny so entertaining for the reader and so valuable for the historian. The introduction of the cherry-tree into Italy (**NH,** XV.102); the mechanical corn-harvester in Gaul, pushed by two oxen (**NH,** XVIII.296); eradication of bracken by cutting it down in spring for two years running (**NH,** XVIII.45); the technology of paper-making from papyrus and the types of paper produced (**NH,** XIII.68-89); the price of different grades of pepper in Rome (**NH,** XII.29); fabrication of artificial flowers (**NH,** XXI.5); propagation of a mutant cucumber with quince-shaped fruit (**NH**XIX.67); the ceremonial use of mistletoe by the druids (**NH,** XVI.249): these items are a small random selection which illustrates the wealth to be found.

I will conclude by mentioning just one item of information which happens to be of singular interest as a contribution to the history of botany (**NH,** XXIV.154; XXV.8.27). Pliny remarks that the first to paint pictures of plants - as a means of identification - were Greeks, the famous botanist Crateuas (physician to king Mithridates of Pontus), and two unknowns, Dionysius and Metrodorus. This was the beginning of botanical illustration, intended in the first place to help physicians recognise medicinal plants. Production of illustrated herbals began therefore about 100 BC with Crateuas, and was well established by Pliny's time, for he tells us that he himself saw illustrations of three kinds of **dracontium** (species of **Arum**). He tells us also that moly, the sacred plant of the gods, described by Homer as having shining white flowers, was pictured with yellow flowers.

In these passages Pliny illuminates an obscure but very important stage in the transmission of herbal knowledge. Figures of herbs, derived directly or at short remove from Crateuas, are known to have been associated at a very early date with the **Materia Medica** of Dioscorides. These illustrations from the classical Greek herbal were transmitted with many of the manuscripts of Dioscorides through the middle ages, and reached some of the earliest printed herbals of the renaissance. In spite of degeneration due to successive copying - a hazard mentioned by Pliny - many figures remained recognisable and provided later botanists with visual evidence to help them establish the identity (or non-identity) of their own local plants with plants of the classical herbal. This was an important link in the chain of plant description and nomenclature.

The interesting fact is that early botanical illustration became associated with the plants of Dioscorides and not with those (for the most part the same) of Pliny. I suspect that the main reason for this was that the more professional **Materia Medica** of Dioscorides became the standard guide for doctors, not Pliny´s discursive and ill-arranged compilation. Certainly in botany he was supremely amateur, with the amateur´s limited vision, yet gifted with the precious quixotic, irresistible enthusiam that can transform the amateur into the salt of the earth.

NOTE

1. J Stannard, "Pliny and Roman botany", **Isis**, 56, 1965, pp. 420-425

Chapter Seven

ASPECTS OF PLINY´S ZOOLOGY

L Bodson

Among the disciplines which are covered by the
Natural History, zoology comes second only to
botany.   Indeed, in his systematic account of the
content of the universe, Pliny devotes no less than
nine books to animals.   In books VIII to XI he aims
at studying them from a zoological point of view,
while in books XXVIII to XXXII he examines their
role in pharmacology.   These extensive accounts are
supplemented   by   some   incidental   mentions in the
encyclopedia,   for   instance   in   book   VII   on
anthropology (see the first and following chapters),
in the chapters on agriculture (book XVIII.249-253,
295-300, 361-363 for example) - on the whole several
hundreds of items of data and comments on both wild
and   domesticated   animals considered in all aspects
from anatomy to economy, magic and popular beliefs.
Pliny´s zoological   contribution   exerted,   long
after   antiquity,   a   durable   influence   on   later
writers interested in animals,(1) until it became
nearly   as   neglected   as   it   had   formerly   been
praised.(2)   The bibliography on Pliny´s zoology
shows   today   a   sharp   contrast   between   the   past
success of this part of his work and the relatively
small   amount   of   recent   studies   dealing with it.
Two main causes can be found for this somehow
paradoxical   situation: the   widening   gap   between
ancient and modern science, which concealed for a
while the historical meaning of Plinian zoology, and
the way critical research on Pliny was orientated a
few decades ago.   By the end of the nineteenth and
in   the   early   twentieth   century   the   prevailing
Quellenforschung focused on the nature and origins
of Pliny´s zoological information, more especially
on the Aristotelian biological treatises which are,
as is well known, the major extant source for Pliny
for most of the problems related to animals.(3)   As

it turned out, the result did not always meet the expectations(4) and the Quellenforschung issue subsided but not without biasing for a while the comparison between Aristotle and Pliny. Too often modern scholars praising the Greek philosopher´s contribution to zoology and biology, even to some excess,(5) expected Pliny to do better and more than his illustrious predecessor. Since this is seldom the case, Pliny´s zoology was, apart from some notable exceptions(6) dismissed or at least belittled. Eventually a much needed counter reaction occurred and took into account Pliny´s own declaration on his aims and goals, before attempting any evaluation of his work and of his impact on the zoological sciences.(7) Indeed, more than once Pliny puts forward or alludes to his intentions. Far from boasting that he intends to match or even challenge Aristotle, he makes it plain that he wishes to offer, for the first time in Latin, "a compendium of Aristotle´s famous works on zoology, with the addition of facts unknown to him". (VIII.44) Following the general principle that he mentions about astronomical matters (II.85) that "it is proper to put (the data) forward because they have been put forward already" (**Prodenda qui sunt prodita**), he firmly emphasises his purpose again in his book on insects (XI.8): "to point out the manifest properties of objects, not to search for doubtful causes." Yet, for all his tireless reading and gathering of information,(8) he does not conceal his limitations. In his preface (chapter 28) he says: "For my own part I frankly confess that my works would admit of a great deal of amplification, and not only those now in question but all my publications." Such a remark (which he develops at some length in chapters 28-32) is not a ready-made phrase to forestall criticisms and objections. It expresses before anything else the lucid self-criticism of a man so unremitting in his scholarly duties that he eventually died in harness.(9) Since a more balanced approach has now been adopted in studies on Pliny, judgements on his works have also become more equitable. Yet, however well-founded they may be, they are and will remain general until extensive and co-ordinated research is undertaken on the disciplines he deals with. As a modest contribution to a much needed and promising undertaking, this paper will provide a short survey of Pliny´s zoology including the features related to pharmacology. His sources, methods and aims will be examined throughout some chapters which illustrate the enlarged knowledge of the palaearctic fauna that had been gained by the first century BC and which is reflected in Pliny's encyclopedia.

Books VIII to XI are successively devoted to "animals living on the earth", that is, mammals and reptiles, sea- and freshwater species, birds and insects. Pliny is here indebted to many authors and mainly to Aristotle.(10) Among other divergences from his prime model, the most immediate difference lies in the way Pliny organises and sets out his material. As has often been pointed out,(11) he does not look for any comprehensive classification of the animal kingdom nor does he even try to set up any systematic description of the characters of differentiation. According to his own method of investigation recorded in his nephew's letter,(12) he adopts the same general, though never compulsory, plan for each book on animals. He merely starts with the largest of the groups under consideration: elephants (VIII.1 et seq.), whales (IX.4 et seq.), ostriches (X.1 et seq.), and then he turns to "creatures of immeasurably minute structure", that is, bees (XI.1 et seq.) as being "the chief insect species". From then on he proceeds to the other species introducing, whenever he judges it suitable for his purpose, digressions, comments and parallels. Book X gives a good example of his manner. Along with chapters on bird and flying mammal anatomy and physiology,(13) most of them imitated or translated from Aristotle, it includes some data on the role of birds in the art of divination,(14) on the first aviaries built in Rome,(15) and on the luxurious taste the Romans of the first century had developed for rare and exotic birds(16) - a habit that Pliny criticises vigorously more than once.(17) The book ends with an account of animal reproduction, sense, nutrition and behavioural interactions (chapters 169-212), mainly inspired by Aristotle's **History of Animals**, V-VI. Similar remarks may be made about books XXVIII-XXXII in which his attention turns from the animals to their parts and products, the properties of which are proved or supposed to be able to cure or relieve illnesses, diseases and physical accidents. Even at the risk of causing disgust, and for the sake and benefit of human life, as he puts it in book XXVIII.2, Pliny enumerates and comments upon all kinds of health troubles and remedies. This leads him to develop his description of some animals which endanger human life, such as parasites and venomous

species.   He   also   introduces   further   zoological
details of some animals mentioned either earlier or
for the first time which are of special interest due
to   their   particular   use   in   medicine.   His
statements on **Naja nigricollis**, the   spitting   cobra
(XXVIII.65), on corals (XXXII.21-24), on fresh-water
turtles   (XXXII.32) and so on, are a few examples of
complementary zoological   material   that   Pliny´s
section on animals offers to the reader.   Yet in
these books, as in the previous ones, puzzling
questions often arise from the haziness of the
descriptions or from evident inaccuracies which make
it impossible or very difficult to identify the
animal in question.   The problem of determining the
animal´s genus or species is of consequence at two
levels: first in the interpretation of the passages
themselves and secondly in the attempt to discover
how many different animals are mentioned by Pliny,
especially in comparison with Aristotle´s figures.
Incidentally, all the ancient descriptions of
animals,   including   Aristotle´s,   raise   similar
questions of identification.   After   Leitner´s
work,(18) which still relies too strictly on a
lexicological approach, progress may be expected
from any further research which takes into account
all the available criteria instead of being confined
to the misleading similarities between the ancient
names   in   Greek   and   Latin   and   the   binominal
terminology coined by modern taxonomists.(19)   But
even if the remaining uncertainties were brought
down to the minimum, there would still remain an
irreducible measure of approximation.   With this in
mind, one finds then in Steier´s chart a rough, yet
instructive guide to the number and distribution of
animals in Aristotle´s and Pliny´s works.(20)   For
the range, accuracy and interpretation of biological
data, Aristotle comes first.   Pliny   enlarges
Aristotle´s recording of animals (either species,
genus or family, according to modern terminology) by
more than 150 units.   The extension of the world
known at the time and the exploration of the
remotest parts of the Roman empire explain these
figures.   They should not be minimized, despite the
errors which mar the text more than once and prevent
it being correctly interpreted.   They even take on
a fuller meaning when one analyses them in the
general context of Pliny´s most reliable sources of
zoological information.   Besides the "Magi" and
anonymous and popular hearsay, Pliny relies on
sources that may be classified as follows:

1. Authors mentioned by name whose works are preserved, thereby generally allowing an easy comparison of the original with Pliny´s version(21).

2. Authors not mentioned but who, however, may be identified(22).

3. Authors mentioned by name whose works are lost(23).

4. No reference given. In this case the problem is to determine whether Pliny is likely to be offering a personal observation or is echoing information of anonymous origin.

In the last two cases, Pliny provides more than once the first and even the only record in antiquity of species unknown to Aristotle and his successors. His testimony gives a unique, though not always the best statement on animals the Romans had become acquainted with for the different reasons clearly characterised by Toynbee.(24) As an example of Pliny´s original contribution to the description of the palaearctic fauna, his account of the Alpine species he knew of will now be examined.

Although the tribes settled in the Alps did not submit to Rome before the turn of the first century AD,(25) the area had come much earlier to Rome´s attention. This was due to the trade established early with the Transalpine side,(26) to the unrest of the inhabitants of the Cisalpine side and the threat they represented for the central parts of Italy,(27) and finally to Hannibal´s expedition in the second Punic war.(28) Strabo mentions wild horses and cattle in the Alps(29) and he quotes Polybius´ description of a wild, deer-like mammal, most probably the European elk (**Alces alces**) which was still present in the Alps at the end of the Roman period, a survivor from the last glaciation.(30) The mountain hare alluded to by Aristotle(31) is mentioned as present in the Alps by Varro.(32) All the other ancient evidence is to be found in Pliny´s **Natural History.** Four wild mammals: chamois (**Rupicapra rupicapra**),(33) ibex (**Capra ibex**),(34) mountain hare (**Lepus timidus**),(35) and marmot (**marmota marmata**,(36) besides goats and cattle;(37) four native birds: ptarmigan (**Lagopus mutus**),(38) black grouse (**Lyrurus tetrix**) and capercaillie (**Tetrao urogallus**),(39) Alpine chough

(Pyrrhocorax graculus),(40) and a migratory species from Egypt, either **Plegadis falcinellus** (glossy ibis), or perhaps the famous **Gerontius eremita** (bald ibis),(41); one fish: burbot (**Lota lota**),(42) and one species of snail(43) are listed. Most of them are so exactly described, at least in regard to their most meaningful characteristics, that the identification of the species can be made beyond all doubt.

Except for Egnatius Calvinus, **praefectus in Alpibus** (X.134), to whom he refers about the ibis, Pliny does not mention any source. Egnatius Calvinus remains otherwise unknown. All that may be said of him is deduced from his title. The charge of **praefectus** for the Alps was created by Augustus after the pacification of the area, which he achieved in 7 BC, and lasted until about 63 AD, when Nero replaced it by a procuratorship.(44) Egnatius Calvinus is not listed by Pliny in his "Table of contents and authorities" (book I) since this table includes the written sources only. For the other features, Pliny may well have collected some of them, as suggested by Steier,(45) while on military duty from about 46 to about 53/54 or 57/58 AD.(46) His statements about the ibex, marmot, mountain hare, ptarmigan, capercaillie and black grouse are so accurately and vividly written that they sound like eyewitness reports, supported by personal and careful observation in the field. They are likely to have been completed with further information collected in Rome where several of these species were imported either for the circus parades and games,(47) or for the aviaries,(48) or as delicacies for some imperial banquet.(49)

Yet, while naming one of these animals by what must be considered as a dialectical term, **ibex**,(50) Pliny uses three Greek names: **tetrao, pyrrhocroax, lagopus**. They are likely to have been taken from a ´field guide´ by some Greek author, perhaps a citizen from Massalia,(51) who had translated and adapted the vernacular animal names into the Greek. Whatever his sources may be, Pliny proves himself to be both well informed and well inspired in his account, which has become of greater historical importance since the animals he mentions have over the centuries become endangered and even threatened with extinction.(52)

When considering the wide range of the Alpine fauna, one cannot however help wondering about the limits of Pliny´s list.(53) The answer to this question should take into account the initial remark

of the **Natural History**´s preface (28), mentioned
above.(54) But there are two other factors not to
be underestimated. The first one is inferred from
his dealing with the fauna of Germany. While on the
limes of the Rhine, Pliny was granted the best
opportunity to get first-hand information on the
animals, especially on the mammals, of the country.
However, he starts his main chapter with this
puzzling remark:"Germany produces few animals".(55)
Then he goes on with another statement of historical
importance on "the most remarkable breeds of wild
oxen", namely European bison and aurochs, two
species later endangered by extinction.(56) The
argument **ex silentio** often proves to be tricky in
use. Yet in this case there is no doubt that Pliny
knew well of the wolves, bears, boars, deer and so
on of Germany.(57) But they were common enough in
other regions not to require any further or special
mention.(58) Then, also, he focuses only on the
most typical species which are also best suited to
complete Aristotle´s descriptions. (59) This also
seems to be true about the Alpine fauna. The few
species which are described or pointed out are among
the most remarkable of the area considered and are
therefore preferred. Typical as they are of the
Alpine fauna, the species occurring in the **Natural
History** are however a very restricted sample of the
animals of the area. Even the least demanding
reader cannot help wondering about the reasons for
such a limited choice, which gives no place to
insects, reptiles and fish, while it omits a great
number of birds and mammals.(60) This feature
leads us to the second point to be stressed.
Pliny´s Alpine fauna is exclusively composed of game
animals and game birds, as may be seen from a
comparison with modern treatises, such as, for
example, **Le gibier des montagnes francaises**(61), in
which there are as many chapters as Alpine animals
in the **Natural History**, each chapter corresponding
to one of the animals described or mentioned by
Pliny for the area considered. Therefore, either
by choice or under his sources´ influence, Pliny
tightly combines two different standpoints - one
rather theoretical, the other more practical - even
in this part of the encyclopedia (books VII-XI) in
which he is usually perceived as a compiler of
zoological data, in contrast to books XXVIII-XXXII,
in which the topic itself commands him to be more
practical-minded. The process is not exclusive.
Yet few, if any, Latin authors illustrate it more
extensively than Pliny, who reflects the pragmatical

interest that the Romans took in animals.

Limited as it is, Pliny´s description of the Alpine fauna turns out to be a representative example of the **Natural History**´s zoological content. Neither the Latin eqivalent of - **mutatis mutandis** - the **Guinness Book of Animal Records**(62) nor a mere reference book in ancient animal lore, Pliny´s work conveys a mixture of both scientific and non scientific information on animals. It only remains to sort it out in all aspects so as to have eventually Pliny´s and the Romans´ contributions to the zoological sciences appreciated at their real value.

NOTES

I wish to thank Dr André Moulin (University of Liège, Dept. English Language, for his friendly help in revising the English version of this paper.

1.   E W Gudger, "Pliny´s Historia Naturalis: the most popular natural history ever published", **Isis**, 6, 1924, pp. 269-281. M Chibnall, "Pliny´s Natural History and the middle ages", in T A Dorey, ed., **Empire and Aftermath. Silver Latin**, II, London, Boston, Routledge and Kegan Paul, 1975 pp. 57-78
2.   See V Carus, **Histoire de la zoologie depuis l´antiquité jusqu´au XIXe siècle**, trans. P -O Hagenmuller, Paris, J -B Baillière, 1880, pp. 68-70; P Brunet and A Miele, **Histoire des Sciences. Antiquité**, Paris, Payot, 1935, pp. 689-690; C Singer, **A History of Biology to about the Year 1900**, 3rd ed., London, New York, Abelard - Schuman, 1959, pp.58-60
3.   See H Le Bonniec, **Bibliographie de l´Histoire naturelle de Pline l´Ancien**, Paris, Les Belles Lettres, 1946, pp. 11-12, 33-35, 43; K Ziegler, s v **Plinius (5)**, in **Real-Encyclopädie**,XXI, 1 (1951), col. 271-274; W Kroll, ibid., col. 309-319, 424-426
4.   D E Eichholz, "Pliny the elder", in **The Oxford Classical Dictionary**, 2nd ed., 1970, pp. 845-6; "Pliny (Gaius Plinius Secundus)", in **Dictionary of Scientific Biography**, ed. C Gillespie, New York, Charles Scribner´s Sons, 1975, vol. XI, pp. 38-40
5.   S Byl, **Recherches sur les grandes traités biologiques d´Aristote: sources ecrites et préjugés**, Brussels, Palais des Academies, 1980, pp. vii-xxix (although Byl underrates Aristotle more than once).
6.   A good example is given by J Scarborough,

"Some beetles in Pliny´s Natural History",
Coleopterists Bulletin, 31, 1977, pp. 293-296
7. H Rackham, **Pliny. Natural History**, London,
Cambridge, Mass., Heinemann, Harvard University
Press, 1938-1963, 10 vols.: vol.I, pp. viii-xi;
introduction at book VIII (A Ernout, 1952), at book
IX (E de Saint-Denis, 1955), at book XI (A Ernout,
1947), in the French edition, Paris, Les Belles
Lettres (Coll. des Universites de France).
8. Pliny the Younger, **Letters**, III.5.7-20
9. Ibid., VI.16. 18-20
10. **Natural History**, I   (table of contents
and authorities), on books VIII-Xi
11. See above, nn. 4,7
12. See above, n.8
13. **Natural History**, X.111-114, 143-153, 168
14. Ibid., 49, 154
15. Ibid.,110,141
16. Ibid., 52, 139-142, 145
17. Ibid., 142
18. H Leitner, **Zoologisches Terminologie beim
Älteren Plinius**, Hildesheim, H A Gerstenberg, 1972
19. L Bodson, "Les Grecs et leurs serpents.
Premiers résultats de l´étude taxonomique des
sources anciennes", in **L´Antiquité classique**, 50,
1981, pp. 57-58; "L´incubation bucco-pharyngienne de
Sarotherodon niloticus (Pisces; Cichlidae) dans la
tradition grecque ancienne", **Archives internationales d´histoire des sciences**, 31, 1981, pp.
5-24.
20. A Steier, **Aristoteles und Plinius.
Studien zur Geschichte der Zoologie**, Würzburg,
Kabitzch, 1913, p. 113 (= **Die Tierformen des
Plinius**, 1912, p. 65) On Steier´s position, see
Kroll, n.3 above, p. 309
21. E.g. VIII.167-168: Varro, **De Re Rustica**,
II.6-8
22. E.g. XI.11-70: Aristotle, **History of
Animals**, IX.40
23. E.g. IX.40: Cassius Dionysius, Roman
translator of Mago´s **On Agriculture**.
24. J M C Toynbee, **Animals in Roman Life and
Art**, London, Thames and Hudson, 1973, pp. 15-23.
25. C M Wells, **The German Policy of Augustus.
An Examination of the Archaeological Evidence**.
Oxford, Clarendon Press, 1972, pp. 59-89
26. C E Stevens, "Gaul (Cisalpine)", in **Oxf.
Class. Dict.**, 2nd ed., Oxford, Clarendon Press,
1970, pp. 458-459. D van Berchem, **Les routes et
l´histoire. Etudes sur les Helvètes et leur
voisins dans l´empire romain**, Geneva, Droz, 1982,

pp. 79-85, 185-217

27.   E T Salmon,   "Cisalpine Gaul" in O.C.D.,
pp. 242-3; C Peyre, La cisalpine gauloise du IIIe au
Ier siècle avant J -C, Paris, Presses de l´Ecole
Normale Superieure , 1979, pp. 43-52

28.   Van Berchem, n. 26 above, pp. 189-190; M
R Sauter, Switzerland from the Earliest Times to the
Roman Conquest, London, Thames and Hudson, 1976, pp.
137-145

29.   Strabo, IV.6.10 (207C)

30.   Id., IV.6.10 (207-208 C); Polybius, XXXIV.
10.8-9.   H Hartmann-Frick, ´Die Tierwelt   im
neolithischen   Siedlungraum",   in   Ur- und
Frühgeschichteliche Archäologie der Schweiz.   II.
Die jüngere Zeit, Basle, Schweizerische Gesellschaft
für Ur- und Frühgeschichte, 1969, p. 19

31.   Aristotle, On Colours, VI.798 a 26-27

32.   Varro, De Re Rustica, III.12.5-6

33.   Natural History, VIII.214: Sunt rupicapr-
ae, sunt ibices... illa Alpes... mittunt.   Also
XXVIII.231 on Rupicapra in pharmacology.   Compare P
Pomet,   Histoire générale des drogues, Paris, J -B
Loyson, A Pillon - Est. Ducastin, 1964, part II, pp.
39-40.   On rupicapra in the history of zoology: M
Couturier, Le chamois (Rupicapra rupicapra L.,) ,
Grenoble, Arthaud, 1938, pp. 429-450

34.   Natural History, VIII.214: Ibices, per-
nicitatis mirandae, quamquam onerato capite vastis
cornibus gladiorum ceu vaginis.   In haec se librat,
ut tormento aliquo rotatus, in petras potissimum, ex
monte aliquo in alium transilire quaerens, atque
recussu pernicis quo libuit exultat. Cf. M Couturier, Le Bouquetin des Alpes (Capra aegagrus ibex
ibex L.), Grenoble, (edited by the author) 1962, pp.
889-919

35.   Natural History, VIII.217:   Et leporum
plura sunt genera.   In Alpibus candidi, quibus
hibernis mensibus pro cibatu nivem credunt esse;
certe liquescente ea rutilescunt annis omnibus; et
est alioqui animal intolerandi rigoris alumnum.

36.   Natural History, VIII.132: Conduntur at
Alpini (mures) quibus magnitudo melium est, sed hi
pabulo ante in specus convecto, cum quidem narrant
alternos mare, ac feminam subrosae conplexos fascem
herbae supinos, cauda mordicus adprehensa, invicem
detrahi ad specum ideoque illo tempore detrito esse
dorso.   M Couturier, Le gibier des montagnes
francaises, 2nd ed., Grenoble, Arthaud, 1981, pp.
139-140, criticises Pliny´s description of haymaking
by the Marmots, still vivid in the Alps, and shows
the origin of this erroneous interpretation. X.186:

Idem (Aegypti mures) bipedes ambulant ceu Alpini quoque.

37. **Natural History,** VIII.179: **Non** degeneres existimandi etiam minus laudato aspectu: plurimum lactis Alpinis, quibus minimum corporis, plurimum laboris capite, non cervice, iunctis. 214: Sunt caprae.

38. **Natural History,** X.133-134: sicut **Alpium** ...et praecipua sapore lagopus; pedes leporini villo nomen hoc dedere cetero candidae, columbarum magnitudine. Non extra terram eam vesci facile, quando nec vita mansuescit et corpus ocissime marcescit.

39. **Natural History,** X.56-57: Decet tetraonas suus nitor absolutaque nigritia, in superciliis cocci rubor. Alterum eorum genus vulturum magnitudinem excedit, quorum et colorem reddit, nec ulla ales, excepto struthocamelo, maius corpore inplens pondus, in tantum aucta uti in terra quoque immobilis prehendatur. Gignunt eos Alpes et septentrionalis regio. In aviariis saporem perdunt. Moriuntur contumacia spiritu revocato.

40. **Natural History,** X.133: sicut **Alpium** pyrrhocorax, luteo rostro niger. On these species (nn.38-40): P Lebreton, **Atlas ornithologique Rhone-Alpes,** Lyon, Centre ornithologique, 1977, pp. 114 (ptarmigan), 114, 116 (black grouse and capercaillie), 282 (Alpine chough).

41. **Natural History,** X.134: **Visam in Alpibus ab se peculiarem Aegypti et ibim Egnatius Calvinus praefectus earum prodidit.** On **Plegadis falcinellus** in Switzerland: P Geroudet, **Grands Echassiers, Gallinacés, Râles d´Europe,** Neuchatêl, Lausanne, Paris, Delachaux-Niestlé, 1987, p. 141; on **Gerontius eremita,** described by C Gesner, **Historia Animalium,** vol. II, Frankfurt, 1617, pp. 309-310 (Waldrapp), see Geroudet, p. 142

42. **Natural History,** IX.63: mirum dictu, inter Alpis quoque lacus Raetiae Brigantinus aemulas marinis (mustelis) generat. H J Cotte, **Poissons et animaux aquatiques au temps de Pline. Commentaires sur le livre IX de l´Histoire naturelle de Pline,** Gap, Loius Jean, 1944, pp. 132-133; E de Saint-Denis, **Le vocabulaire des animaux marins en latin classique,** Paris, Klincksieck, 1948, pp. 73-74

43. **Natural History,** VIII.140: Est aliud genus (coclearum) minus vulgare, adhaerente operculo eiusdem testae se operiens. Obrutae terra semper hae et circa maritimas tantum Alpes quondam effosae, coepere iam erui et in Veliturno.

44. C Jullian, Histoire de la Gaulle. IV. Le

governement de Rome, Paris, Hachette, 1913, pp. 59-60; F Staehelin, **Die Schweiz in römischer Zeit,** 3rd ed., Basle, B Schwabe, 1948, pp. 166-196; J Prieur, **La Province romaine des Alpes cottiennes,** Villeurbanne, R Gauthier, 1968, pp. 83-85, 130

45. Steier, (n.20 above) pp. 57-58 (=pp. 9-10)

46. J Beaujeu, **Pline l´Ancien. Histoire naturelle,** I, Paris, Belles Lettres, 1950, p. 7

47. O Keller, **Die antike Tierwelt,** I, Leipzig, J Cramer, 1909, pp. 183-4, 212-3, 299; G Jennison, **Animals for Show and Pleasure in Ancient Rome,** Manchester, Manchester University Press, 1937, pp. 89, 93. 131; Toynbee, no 24 above, pp. 18-19, 25, 147, 287

48. Jennison, pp. 107, 114; Toynbee, p. 256

49. Suetonius, **Cal.,** XXII.7; J Andre, **L´alimentation et la cuisine à Rome,** Paris, Klincksieck, 1961, pp. 121-122, 126

50. A Ernout, A Meillet, **Dictionnaire étymologique de la langue latine,** 4th ed., Paris, Klincksieck, 1967, p. 305

51. O Keller, no 47 above, vol.II (Leipzig, J Cramer, 1913, pp. 91-2, 110, 156, 165-6; D´A W Thompson, **A Glossary of Greek Birds,** Oxford, Clarendon Press, 1936, pp. 110, 152, 168; J André, **Les noms d´oiseaux en latin,** Paris, Klinckseick, 1967, pp. 97-8, 135, 151-2; F Capponi, **Ornithologia Latina,** Genoa, Istituto di Filologia classica e medievale, 1979, pp. 311-2, 437-8, 483-4

52. R Hainard, **Mammifères sauvages d´Europe,** vol. II, Neuchâtel, Delachaux et Niestlé, 1962, pp. 122-137, 154-157, 173-186; R P Bille, **Les animaux de montagne,** Paris, Denoel, 1979 p. 144-5, 152-3, 157, 184-5

53. P C Rougeot, **Guide du naturaliste dans les Alpes,** Neuchâtel, Delachaux et Niestlé, 1972,; Bille, n. 52 above.

54. See above, p.10

55. **Natural History,** VIII.38

56. F Zeuner, **A History of Domesticated Animals,** London, Hutchinson, 1963, pp. 201-214

57. Caesar, **De Bello Gallico,** VI.26-28

58. **Natural History,** VIII.80-83, 227; 126-131, 228; 210-213; 112-119, 227, 228

59. See above, p. 11

60. See Bille, n. 52 above.

61. M Couturier, **Le gibier des montagnes françaises,** 2nd ed., Grenoble, Arthaud, 1981, chapters 2-7

62. G L Wood, **Animal Facts and Feats. Thousands of amazing Animal Achievements and Fabul-**

ous Facts fron the Natural World, Toronto, London, New York, Bantam Books, 1977

Chapter Eight

PLINY ON MINERALOGY AND METALS

J F Healy

Introduction
Pliny the Elder, in the preface(1) to his **Natural
History**, explains that the subject of his study, or
´enquiry´ (**historia**)(2) is "a barren one, the world
of nature (**rerum naturae**) that is life". His
treatment of this theme, however, differs markedly
from that of Lucretius in the **De Rerum Natura**(3):
fundamentally it is more scientific than philosophical. Pliny´s specific aim, to save the
science of the past from the forgetful indifference
of the present,(4) is well illustrated in the last
five books of the **Natural History**, in which he
treats topics which come within the wider field of
earth sciences.
    In 73 AD Gaius Plinius Secundus became procurator
of **Hispania Terraconensis**,(5) an appointment of no
small significance for the section of the **Natural
History** (especially in XXXIII) devoted to gold
mining since it enabled him to gain first hand
knowledge of mining operations in Asturia and
Gallaecia(6) and of the technology involved in
processing gold. This is confirmed by the accuracy
and detail of his account and by the use of local
Spanish and other mining terms - including Greek:
**aut rusticis vocabulis aut externis immo barbaris
etiam, cum honoris praefatione ponendis.**(7)
    In addition, books XXXIII-XXXVII embrace Pliny´s
general interest in minerals, ore minerals, metals,
stones, earths and gemstones. A close examination
of the text of the **Natural History** not only shows
Pliny´s direct debt to earlier authorities, many of
whose works are lost, but reveals, in addition, his
own considerable contribution to mineralogy and
related topics. A **caveat** however, is necessary at
this point namely to the effect that the value of
all ancient writers on science and technology should

be assessed in the light of the accuracy of their observations rather than by the criterion of analytical research on their part. Thus Pliny was able to describe the crystal-systems of quartz,(8) diamonds,(9) beryls,(10) and the ´rainbow´ stone (**iris**),(11) with an acceptable degree of accuracy, although the actual science of crystallography was not established until the nineteenth century with the works of Romé de l´Isle(12) and abbé Haüy.(13) Similarly the pyroelectric properties of amber were known to Theophrastos(14) and, additionally, of tourmaline, to Pliny,(15) but detailed investigation of the phenomenon of static electricity had to wait until William Gilbert´s experiments in the sixteenth century.(16)

The Greeks and Romans had only a limited knowledge of geology(17): they were mainly interested in the origin and source of ore minerals, the metals derived from them, earths, pigments, building materials and precious stones, that is, in economic geology and petrology. Their approach was descriptive and neither race developed a classification based on the origin and gradual transformation of rocks. Aristotle is exceptional in so far as his theory of the production of minerals from vaporous exhalations(18) comes very near to being a description of the process known as **pneumatolysis**,(19) by which vapours or heated waters discharged from igneous magmas by volcanic action are deposited on cooling in rock fissures, to form mineral deposits.

Greek and Roman theories about the formation of minerals
Natural science began with the Ionian philosophers, or Monists,(20) who were the first to try to determine the nature of the universe and man´s place in the wider order of things. They shared a common belief in the existence of four basic elements, fire, air, water and earth, each choosing **one** as the basis of the physical universe. Thales(21) (c.585 BC) suggested water, a logical choice, since it could be observed in the three states of matter, as liquid, solid and vapour.

Empedocles(22) (492-432 BC) regarded the universe as being built up of the four traditional elements, by the varied interaction of which world changes came about. He sought to uphold his theories by observation, being the first to state that the earth´s interior was in a molten condition, because molten material was intermittently thrown up by mount Etna.

In the field of mineralogy, however, the pre-Socratics added little of consequence. Indeed, the nature and composition of physical substances had not attracted the interest of these early ´scientists´ (**physiologoi**). After Parmenides(23) (c.450 BC) the concept of mixture and association replaced that of **genesis**.

Plato(24) (c.429-347 BC) was the first to discuss the formation of rocks in any detail; Aristotle(25) (384-322 BC) in part agreeing with Plato, defined

> ...two exhalations - one vaporous, the other smoky, and there are two corresponding kinds of body produced within the earth, namely **fossiles** (quarried materials) and **metals**. The dry exhalation, by the action of its heat, produces all the **fossiles**; for example realgar (a red disulphide of arsenic), ochre, ruddle (a red variety of ochre) and sulphur... Metals are the product of the vaporous exhalation.(26)

In Aristotle´s account, however, the part played by the dry exhalation in producing the **fossiles** is obscure. Eichholz concludes that the fire and heat from this form the **fossiles** by reducing earth to the consistency of fire ash and perhaps causing it (though this is less certain) to take on bright colours. Those of the **fossiles** which are stones must, furthermore, have been hardened by this heat.(27) The conclusion that the dry exhalation is the efficient and not the material cause of Aristotle´s **fossiles** seems to be corroborated by Theophrastos(28) who discusses the origin of a group of mineral earths, of which three, namely ochre, ruddle and realgar, are cited by Aristotle(29) as **fossiles**. "Some of them seem to have been exposed to fire (**pepyromena**) and burnt (**katakekaymena**), such as realgar, orpiment (yellow trisulphide of arsenic) and others of the same kind. In short, all of these result from a dry and smoky exhalation." The passage makes sense only if the dry exhalation is the efficient cause of the formation of these earths.(30)

Theophrastos (372-287 BC), who was a pupil of Aristotle, wrote the **De Lapidibus**,(31) the oldest extant scientific treatise dealing expressly with minerals. He describes the formation of minerals and metals as follows: (32)

> Of the substances formed in the ground, some
> are made of water and some of earth. The
> metals obtained by mining, such as silver, gold
> and so on, come from water; from earth some
> stones, including the more unusual because of
> their colour, smoothness, density or other
> quality...

and continues

> Some things are solidified through heat, others
> through cold. And probably there is nothing
> to prevent some kinds of stones being formed by
> either of these two methods, although it would
> seem that all the types of earth are produced
> by fire, since things become solid or melt as a
> result of opposite forces.

Aristotle had also stated that solidification of
some substances occurs in both ways, as for example,
mud. Theophrastos attempts to study minerals in a
systematic way, classifying sixteen species - from
the whole range known to the Greeks - on the basis
of Aristotelian principles, as "metals, stones and
earths". The classification, however, is by
superficial characteristics and not according to any
concept of chemical composition. Nevertheless, **De
Lapidibus** is important both in the history of
mineralogy and of chemical technology.(33)
Comparison of Pliny's text(34) with that of
Theophrastos,(35) where they are describing the same
minerals, confirms that Pliny had direct access to
De Lapidibus. Pliny also relied on a number of
authorities, including Posidonius(36) and
Straton,(37) whose works were subsequently lost.
In particular Pliny follows Posidonius' explanatiuon
of the formation of transparent and semi-transparent
stones.(38)

> The raw material of such stones was water,
> possibly impregnated with earthy particles; and
> this liquid was compacted either by cold in the
> atmosphere or by one or other of the two
> exhalations described by Aristotle, colours
> being imparted to the stones by the dry
> exhalations which also hardened them.(39)

Pliny describes the formation of rock-crystal
(quartz):

> A cause contrary to the one mentioned is

114

responsible for creating rock-crystal, for this is hardened by excessively intense freezing. At any rate it is to be found only in places where winter snows freeze most thoroughly; and that it is a kind of ice is certain: the Greeks have named in accordingly (**krystallos**). Rock crystal also comes to us from the East, for that of India is preferred to any other... in Europe excellent rock-crystal occurs in the ranges of the Alps... What is certain is that it is not found in well-watered localities, however cold the district may be, even if it is one where the rivers freeze down to the bed. The inevitable conclusion is that rock-crystal is formed of moisture from the sky falling as pure snow.(40)
Pliny, however, expresses some scepticism about this theory. Seneca(41)(c. 5/4 BC - 65 AD) describes rock-crystal as "rain water containing a very little earthy matter" (**aqua caelestis minimum in se terreni habens**): Diodorus Siculus(42) says "pure water".

The perfectly limpid, colourless rock-crystal (quartz) stands out prominently among all other minerals by reason of its clearness and transparency in which it often surpasses even the diamond, although it is not comparable with the latter in lustre or play of colours. It commonly occurs in fine crystals, the prism faces of which are almost without exception largely developed, so that the habit of the crystals is columnar. Some crystals may be attached to one or both ends of the matrix and show considerable variation in size.(43) Fluor-spar, the substance of the "myrrhine vases" writes Pliny,(44) "is thought to be a liquid which is solidified underground by heat (**umorem sub terra putant calore densari**), while later(45) Aristotle´s dry exhalation is more precisely defined as **caloris anima**; this exhalation is again regarded as responsible for the hardening and coloration of the mineral. Diodorus Siculus does not mention fluor-spar which, during the greater part of Posidonius´ lifetime, was still a rarity in the western world,(46) but Juba, or Xenocrates, or whoever was responsible for the view expressed by Pliny, seems to have been influenced by the general theory of Posidonius. Selenite(47) (**lapis specularis**) is formed "when a liquid... is frozen and petrified by an exhalation of the earth (**terrae quadam anima**. Later, in his disussion of alum, Pliny(48) shows an awareness of the process of crystallisation:

> All alum is produced from water and slime, that is a substance exuded by the earth; this collects naturally in hollows in winter and its maturity by crystallisation is completed by the sunshine of summer; the part of it which separates earliest is whiter in colour.

Observation of the precipitation of alum, together with the appearance of quartz crystals, may have led to the commonly held view in the ancient world that minerals were in a constant process of growth or regeneration. Thus Pliny(49) writes:

> Among the many marvels of Italy itself is one for which the accomplished natural scientist Papirius Fabianus (the younger Seneca was a pupil of his) vouches, namely that marble actually grows in the quarries; and that the quarrymen, moreover, assert that the scars on the mountain sides fill up of their own accord.

Elsewhere, Pliny,(50) writing of black lead, states

> Black lead is excavated with considerable labour in Spain and through the whole of the Gallic provinces, but in Britain it is found in the surface stratum of the earth... It is remarkable that in the case of these mines only that when they have been abandoned they replenish themselves and become more productive. This seems to be due to air infusing itself to saturation through the open orifices, just as a miscarriage seems to make some women more prolific.

The idea of the regeneration of stone is also cited in Strabo(51): "This is not the only remarkable thing about Elba, there is also the fact that the diggings which have been mined are in time filled up again, as is said to be the case with the ledges of rock in Rhodes and elsewhere."

Similar beliefs have endured in India until the present day, where diamonds are thought to replenish themselves as follows(52):

> Diamantiferous sandstone, which has been moved from its natural bed, and from which the diamonds have been extricated, is often allowed to be exposed to the various atmospheric weathering agencies for some time, and is then again worked over, when a further yield of

of diamonds may be given, this being sometimes repeated several times. This fact has given rise to the belief among the natives that this second crop of diamonds has originated in the waste rock, or that it is the result of a fusion together of the smaller diamonds originally left behind; similar beliefs are also met with in South Africa.

The diamonds have, of course, been released by weathering of larger fragments of rock in which they may have been embedded.

Minerals
(i) Definition
Minerals, by modern definition, are naturally formed, homogeneous, almost exclusively solid constituents of the earth and of extra-terrestrial bodies; they possess definite, but not necessarily fixed or constant chemical composition and are either uncombined elements in a native state, or compounds of elements formed in accordance with chemical laws. Minerals are the result of a sequence of complex processes which ended, in many cases, with crystallisation in rocks or ore bodies, controlled by trivial local factors. They have a characteristic atomic structure which is expressed in their external crystalline form together with other distinctive qualities.

(ii) Terminology
Some of the problems encountered in the interpretation of Pliny´s text clearly derive from textual errors arising out of the transmission of the manuscripts, which is understandable in the light of the subject matter; equally many difficulties and inconsistencies are the result of the speed at which Pliny wrote the **Natural History** and the fact that the work never received a final revision. Pliny the Younger(53) expresses surprise: "You may wonder how so busy a man as my uncle was able to complete so many volumes, many of them involving detailed study". He attributes this to Pliny´s penetrating intellect and amazing powers of concentration... also to the fact that he managed with a minimum of sleep.

The nomenclature of minerals is an area in which Pliny, like most ancient authorities, is especially prone to inconsistency. The common terminations in -ite, -itis, -ites, which had been used by the Greeks, were adopted by the Romans, generally

Mineralogy and Metals

without problems. Names often indicate some
characteristic constituent or use of the mineral, or
the locality where it is chiefly found. So
h(a)ematitite(54) is named after its blood-red
colour (**haima**). Pliny(55) lists other methods of
identification: some stones are named after parts of
the body, as **hepatite** (**hepar**) reflecting the hepatic
appearance of the material, or after an animal
substance - fat - **steatite** (**stear), or an animal´s
colour, as **carcinias (karkinios,** crab). Rock
crystal is named after ice (**krystallos).(56)**
Inanimate objects sometimes provided a name, for
example, **cenchrites** is a stone which looks as if it
had been sprinkled with grains of millet (**kegchros**):
cenchros may, however, also refer to a small
diamond.(57) Some minerals have two or more names:
thus asbestos is known as **asbestos**(58) and as
amiantus.(59) Some terms cover a variety of
substances: **galena** (or **molybdaena**) is generally lead
sulphide, or argentiferous galena,(60) but can be
applied to crude lead.(61) **Silex**(62) is any hard
stone, including Italian tufas and marble. **Lychnis**
and **lychnites**(63) (comapare Strabo)(64) is both a
precious stone of red colour, possibly ruby, and
Parian marble(65) (**lychnites lithos**) because,
according to Varro, it is quarried in galleries by
lamplight. **Magnes lapis**(66) is identified with
magnesian limestone (dolomite) by Bailey,(67) rather
than with magnetite. **Schistos** embraces talc.(68)
Zehnacker(69) suggests that **schistos** is a generic
term for all minerals of lammellar structure.
Anthracites(70) is a mixture of limonite and
magnetite,(71) or of limonite and potash alum.(72)
Finally, **adamas**(73) is used to cover a large number
of hard substances, both mineral and metals,
including diamond,(74) possibly rock-crystal,(75)
iron pyrites,(76) and even platinum,(77) although
the latter is unlikely. Additional difficulty is
caused by Pliny´s practice of postponing any mention
of identifying characteristics until the second or
later appearance of a term in the text, as in the
case of **andromas** (iron pyrites).(78)

A further problem arises from the fact that
modern nomenclature sometimes differs from ancient.
Among a number of examples, **chrysocolla**(79) was used
by the Greeks and Romans as a generic term referring
to any bright green copper mineral. Theo-
phrastos(80) and other authorities apply this name
to any material used in soldering gold. The
mineral currently known as **chrysocolla**, however, is
a hydrous copper silicate which occurs as a

decomposition product of copper ores: it is found as encrusting and botryoidal masses.(81)

(iii) Physical characteristics

Minerals have a number of properties which are derived from their chemical composition or crystalline structure.

(a) Colour

Theophrastos and other writers before Pliny refer mainly to optical characteristics, especially transparency (or opacity), lustre and colour. In the search for ore minerals colour was an all important factor,(82) as in the case of the varieties of magnetite(83) which are thus distinguished. Colour may be related directly to one of the mineral´s major constituent elements and therefore constant and characteristic. Such minerals are known as **idiochromatic**(84) and colour serves as a primary means of identification. For example malachite is green and azurite blue (both forms of copper carbonate), while rhodonite and rhodochrosite are red or pink. The Median(85) stones described by Pliny are malachite intergrown with azurite - a common phenomenon.

In **allochromatic** minerals the colour is attributed to appreciable amounts of an element, such as iron, which has strong pigmenting power. So in sphalerite (zinc sulphide) the progressive substitution of iron for zinc changes the colour from white through yellow and brown to black. Ore minerals, although of constant colour, may show alteration on exposure to air and produce a tarnish different from their original true colour: this is particularly noticeable in copper minerals. Bornite, for example, found in association with other forms of copper in primary deposits, is called ´peacock ore´ because of this phenomenon; it acquires a blue-violet surface film when exposed to air. Copper pyrites is similarly affected.

In the **Natural History** Pliny shows evidence of a much wider understanding of the whole range of physical properties of minerals,(86) among them cleavage, hardness, tenacity, specific gravity, magnetic and electrical characteristics and streak, in addition to optical properties.

(b) Cleavage

A mineral has cleavage if it breaks along definite plane surfaces. Pliny(87) describes the sarc-ophagus stone (**lapis sarcophagus**), which has this property: "at Assos in the Troad we find the sarcophagus stone which splits along a line of cleavage". The best known example of cleavage, however, occurs in selenite (**lapis specularis**)(88): "This has a far more amenable character which allows it to be split into plates as thin as may be wished". For this reason selenite was sometimes used as a substitute for window glass, a practice which continued as late as the eighteenth century.

(c) Hardness

Duritia is alluded to in fluor-spar,(89) onyx marble (**chernites**)(90) limestone (**porus**)(91) and other minerals. This property, in addition to tenacity, the resistance which a mineral offers to breaking crushing and bending, in short its cohesiveness, is discussed in connection with diamonds, Likewise **molybdaena** is described as friable (92) (**friabilis**), that is, ´easily crumbled´. At the other extreme Pliny mentions the softness (**mollitia**) of Eretrian earth.(93)

Tenacity

Tenacity is further defined by the following terms(94): brittle, malleable, sectile, ductile, flexible and elastic, some of which properties are known to Pliny. Jet (**gagates lapis**)(96) is brittle (**fragilis**). Likewise the emeralds from Chalcedon(96) are noted for their brittleness (**fragilitas**). Pliny(97) recognises **malleability** (**facilitas**), observing that no other material is more malleable than gold (**laxius dilatatur**).

(e) Specific gravity

Pliny makes no reference to the specific gravity,(98) or density of minerals, or metals, but states(99) that "Tin is also found in gold mines called **alutiae**, through which a stream of water is passed which washes out the black pebbles of tin (cassiterite) mottled with white spots and of the same weight as the gold and consequently they remain with the gold in the bowls (or baskets) in which it is collected." The specific gravities(100) of tin and gold, however, are 7.3 and 19.3 respectively and what Pliny fails to realise is that both remain because they are relatively more dense than the accompanying gangue materials: quartz, for example has a specific gravity of 2.65.(101) Their mass is by no means comparable. Pliny draws the wrong conclusion elsewhere in his comparison of gold and lead.(102) Plato,(103) by contrast, understands that gold has a very high density (**pyknotes**).

(f) **Magnetic and electrical properties**
Pliny(104) describes the magnetic properties of magnetite and the pyroelectric property of amber(105) and tourmaline.(106)

(g) **Streak**
The colour of a finely powdered mineral is known as its streak,(107) **sucus**. Although the colour of a mineral may vary (for example by tarnish),(108) its streak is usually consistent and thus a useful factor in establishing positive identification. Pliny(109) draws attention to this characteristic: "Eretrian earth is tested (by its softness and) by the fact that it leaves a violet tint (**violacium**) if rubbed on copper" (here acting as a streak plate). Similarly the test for **andromas**(110) (specular iron ore) is to rub it on a whetstone of slate (**lapis basanites**), when, if genuine, it gives off a blood red streak (**sucus**).(111) The use of the touch-stone(112) (**lapis Lydius**) in testing the composition of gold, yields ´streaks´ when a scraping is taken from an **ore**. Finally Pliny(113) mentions the phenomenon of efflorescence: "...belonging to the same stone (**Assos**) is what is called efflorescence (**flos**) which is soft enough to form powder". (Possibly gypsum produced by the action of sulphuric acid, which would have resulted from the decomposition of pyrites, upon the limestone.)(114) Only two major properties (of minerals and metals) are not distinguished, or alluded to, namely fracture and parting.(115) In describing all the other main characteristics of minerals, Pliny makes a significant contribution to the science of mineralogy, antcipating the basis of modern classification.

Technical terminology
While Pliny is careful in his use of technical vocabulary, distinguishing for example between **calefacere** and **coquere**,(116) both terms referring to a heat process, but the latter specifically implying chemical change as a result of the applicaton of heat, in his description of minerals he is often less precise. **Nitor** and **fulgor** generally refer to ´lustre´(117) but are sometimes used of ´brilliance´(118) in colour. **Tralucidus** may mean ´transparent´(119) or equally (as **tralucere**)(120) ´translucent´. The adjective **pinguis** in addition to its basic meaning ´rich´ or ´thick´(121) may mean rich (in colour),(122) ´massive´(123) or ´dull´(124) and, as a noun, **pinguitudo** means ´greasy appearance´.(125) Likewise **crassitudo** may be ´opacity´, but

more probably ´thickness´(126) or ´bulk´(127): yet again **crassiores** is ´more opaque´(128) while, in the phrase **crassius nitere** it means ´shine with a duller lustre´.(129)

Crystallography and crystal systems

In his treatment of quartz, diamonds, beryl and the ´rainbow stone´ (**iris**) Pliny enters the field of elementary crystallography.

In the case of rock crystal, or quartz, he writes(130): "Why it is formed with hexagonal faces (**sexangulus lateribus**) cannot readily be explained: and any explanation is complicated by the fact that, on the one hand its terminal points are not symmetrical and that, on the other, its faces are so perfectly smooth that no craftsman could achieve the same effect."

Of the six types of **adamas**(131) listed, only the Indian and, possibly, the Arabian, can be identified with any degree of certainty.

> There is the Indian, which is not formed in gold and has a certain affinity with rock-crystal which it resembles in respect of its transparency and its smooth faces meeting at six corners (**laterum sexangulo levore**). It tapers to a point in two opposite directions and is all the more remarkable because it is like two whorls joined together at their broadest parts. It can be as large even as a hazel nut. Similar to the Indian, only smaller, is the Arabian which is, moreover, formed under similar conditions.

Pliny is here clearly describing an octohedral diamond.

The extraordinary difference in the appearance of diamond and that of other forms of carbon depends solely on the crystallisation of the material and the physical characteristics consequent upon this.

> The diamond is one of the most perfectly crystallised of minerals. Almost every single stone is bounded by more or less regularly developed faces. The faces of diamond crystal differ from those of most other crystallised minerals in that they are, as a rule, much curved and rounded instead of being perfectly plane, as is usually the case. This curvature is due to the mode of growth of the crystal and not to subsequent attrition as might be

thought.(132)

Pliny´s(133) description of beryls, also found in India, but rarely elswhere, is equally interesting. "Many people consider the nature of beryls to be similar to, if not identical with, that of emeralds. All of them are cut by skilled craftsmen to a smooth hexagonal shape (**sexangula figura**)." He is referring to prisms. Later, Pliny(134) reveals that he is not altogether convinced that beryls are produced artificially. "Some people are of the opinion that they are formed from the start as prisms (**angulosos**)", and again observes "The Indians are extraordinarily delighted at the length of beryls (**mire gaudent longitudine eorum**)." The shape of the beryl is the result of natural crystalline growth, the prisms representing the orderly arrangement of atoms within the structure: beryl belongs to the hexagonal system.

Pliny(135) also describes a so-called ´rainbow stone´ (**iris**),

> dug up on an island in the Red Sea 60 miles distant from the city of Berenice. In every other respect it is merely rock-crystal, and is sometimes called **radix crystalli** ´root of crystal´ for this reason. It is known as **iris** in token of its appearance, for when it is struck by the sunlight in a room it casts the appearance and colour of a rainbow on the walls nearby, continually altering its tints and ever causing more and more astonishment because of its extremely changeable effects. It is agreed that it has hexagonal faces like the rock-crystal and some people assert that it has rough faces and unequal angles; and that in full sunlight it scatters the beams that shine upon it, and yet at the same time lights up adjacent objects by projecting a kind of gleam in front of itself. But, as I have said, it does not produce any colours except in a dark place; and even then, the effect is not as though the stone itself contained the colours, but rather as though it were forcing them to rebound from the wall. The best kind is that which produces the spectra that are the largest in size with the closest resemblance to a rainbow. There is also another ´rainbow stone´, the **iritis**,(136) which is similar to the former in every respect except that it is very hard...a stone which is similar in appear-

ance but different in its effects is the so called 'trifle' stone (**leros**), in which there is a white and black streak traversing the rock-crystal.

Pliny discusses the shapes of gemstones(137):

> Concave (**cava**) or convex (**extuberantes**) stones are considered less valuable than those with a plane surface (**aequalibus**). An elongated shape (**figura oblonga**) is the most valuable; then what is called lenticular (**lenticula**) followed by a flat (**epipedos**) and, finally, a round (**rotunda**) shape. Stones with sharp angles (**angulosis**) find the least favour.

Gold

Pliny's description of the occurrence, mining and processing of gold is based on personal knowledge acquired during his procuratorship in Spain.(138) Two passages refer to the location of gold: in the first, Pliny(139) writes: "Gold more than all other metals is found unalloyed in nuggets, or in the form of **detritus**. Whereas all other metals when found in the mines are brought into a finished conditon by means of fire, gold is gold straightaway and has its substance in a perfect state at once, when it is obtained by mining". Secondly(140), he elaborates:

> Gold in our part of the world - not to speak of Indian gold obtained from ants, or the gold dug up by griffins in Scythia (141) - is obtained in three ways, in the **detritus** of rivers, for example in the Tagus (Tago) in Spain, the Padus (Po) in northern Italy, the Hebrus (Maritza) in Thrace, the Pactolus (Sarabat) in Asia Minor and in the Ganges (India). Another method is by sinking shafts and finally it is sought in the fallen debris of mountains.

He vividly describes gold as being found **marmoris glareae inhaerens** where **marmor** is the quartz matrix. The terms used by Pliny include Roman, Spanish and Greek (**rusticis vocabulis aut externis**)(142): thus **ballux**, **balluca** (nuggets), **palae**, **palai**, **palages**, **palacarnas**, **psalacurnas** (lumps of gold), **arrugiae** (open-cut mines), **hagogai** (channels). Other Iberian words recur in the **Lex metalli Vipsac-ensis**,(143) of the principate of Hadrian (117-138 AD), recorded on two bronze plaques discovered in the slag heaps of the Aljustrel mines in Portugal. Pliny is also the first authority to record the association of gold and silver, stating that where the latter is more than 20 percent of the sample, the metal should be referred to as **electrum**.(144)   A number of minerals are used in the refining of gold; the earliest literary reference to refining occurs in Diodorus Siculus(145) (who follows Agatharcides).

> They put the processed mineral by measure and weight into earthen crucibles.   They mix this with a lump of lead according to the mass, lumps of salt, a little tin and barley bran. They put on a closely-fitting lid carefully smearing it with mud and heat it in a furnace for five days and nights continuously;   then they allow the crucibles to cool and find no residual impurities in them;   the gold they recover in a pure state with little wastage.

Diodorus conflates a cupellation and cementation, or ´parting´ process.(146)   Strabo(147) records a sim- ilar technique in which ´styptic earth´ (**stypt- eriodes ge**), possibly alum and sulphates, was used. Pliny(148) writes: "Gold is heated with twice its weight of salt and three times its weight of **misy** (copper pyrites) and again with two portions of salt and one of **schistos**.   Treated in this way it draws poison out, when the other substances have been burnt up with it in an earthenware crucible while it remains pure and uncorrupted."   In the Leiden/ Stockholm papyrus,(149) **misy**, alum and salt are present in equal quantities (four parts).   The production of **misy**(150) confirms its identification:

> Some people have reported that Misy is made by burning ore-mineral in trenches, its fine powder mixing itself with the ash of the pine-wood burnt; but as a matter of fact though obtained through from the material **chalcitis** (that is, copper pyrites in the process of decomposition), it is part of its substance and separated from it by force, the best kind being obtained in the copper factories of Cyprus, its marks being that when broken it sparkles like gold and when it is ground it has a sandy appearance, without earth, unlike **chalcitis**.

Comparison with the term **misy** in Dioscorides(151) further confirms its nature.   **Schistos** is an equal- ly interesting mineral, and Pliny(152) explains: **concreti aluminis unum genus schiston appellant Graeci**. Bailey(153) reasonably deduced that **schistos** is a form of potash alum. Zehnacker(154) is less specific. "Le mot **schistos** peut designer toutes sortes de mineraux a structure lamellaire." The term is certainly not confined to talc.(155)

Touchstone
The use of the touchstone to examine the purity of gold in the ancient world is well attested. Pliny(156) explains:

> With the mention of gold and silver goes a description of the stone called the ´touch- stone´ (**coticula**), formerly according to Theo- phrastos(157) not usually found anywhere but the river Tmolus, but now found in various piaces. Some people call it Heraclian stone and others Lydian. The pieces are of moderate size, not exceeding four inches in length and and two in breadth.

Pliny clearly follows Theophrastos and both stress that a dry smooth surface is necessary to carry out tests. Pliny, however, misunderstands Theophrastos in one particular and incorrecly claims that "When experts have taken a scraping from an **ore**, they can say at once how much gold it contains and how much silver or copper".
   The stone customarily used is siliceous schist (158) of black colour, the best stones being of a uniform deep black, without light coloured veins or spots and of a fine grain. The surface of the stone must be completely flat and matt, otherwise the touch (streak) will not adhere to the stone.

Silver
Pliny(159) next describes the production of silver:

> silver is only found in deep shafts, and raises no hopes of its existence by any signs, giving off no shining sparkles such as are seen in the case of gold. The ore is sometimes red, some- times ash coloured. It cannot be smelted except when combined with lead or with the vein of lead (**vena plumbi**), called **galena**, lead ore, which is usually found running near veins of silver ore. Also when submitted to the same process of firing, part of the ore precipitates as lead, while the silver floats on the surface like oil on water...The same mines(160) also produce the mineral called ´scum´ of silver (**spuma argenti**). Of this

there are three kinds, with Greek names meaning respectively golden (**chrysitis**), silvery (**argyritis**) and leaden (**molybditis**): and for the most part these colours are found in the same ingots. The Attic kind is the most approved, next the Spanish. The golden scum is obtained from the actual vein, the silvery from silver, and the leaden from smelting the actual lead, which is done at Pozzuoli from which place it takes its name (**argyritis Puteolana**). Each kind, however, is made by heating its raw material till it melts, when it flows down from an upper vessel into a lower one and is lifted out of that with small iron spits and then twisted round on a spit in the actual flame, in order to make it of moderate weight. Really, as may be inferred from its name, it is the scum of a subtance in a state of fusion and in the process of production. It differs from dross in the way in which the scum of a liquid may differ from the lees, one being a blemish excreted by the material when purified. Some people make two classes of scum of silver which they call **scirerytis** and **peumene** (or **reumene**) and a third **molybdaena** leaden scum which we shall speak of under the heading of lead. To make the scum available for use it is boiled a second time after the ingots have been broken into pieces the size of finger-rings. Thus after being heated up with bellows to separate the cinders and ashes from it, it is washed with vinegar or wine and cooled down in the process. In the case of the silver kind in order to give it brilliance the instructions are to break it into pieces the size of a bean and boil it in water in an earthenware pot with the addition of wheat and barley wrapped in new linen cloths, until the silvery scum is cleared of impurities. Afterwards they grind it in mortars for six days, three times daily washing it with cold water and, when they have ceased operations, with hot, and adding salt from a salt-mine, an obol weight to a pound of scum. Then on the last day they store it in a lead vessel. Some boil it with white beans and pearl-barley and dry it in the sun, and others boil it with beans in a white woollen cloth till it ceases to discolour the wool; and then add salt from a salt mine, changing the water from time to time, and put it out to dry on the 40 hottest

127

days of summer. They also boil it in a sow´s
paunch in water and when they take it out, rub
it with soda (**nitro**), and grind it in mortars
with salt as above. In some cases people do
not boil it but grind it up with salt and then
add water to rinse it... It is used to make an
eye-wash and for womens´ skins to remove ugly
scars and spots and as a hair-wash.

Lead
Pliny(161) discusses lead:

There are two different sources of black lead
(**plumbum nigrum**) as it is either found in a
vein of its own, and produces no other substance mixed with it, or it forms together with
silver and is smelted with the two veins mixed
together. Of this substance the liquid which
melts first in the furnaces is called **stagnum**;
the second liquid is argentiferous lead, and
the residue left in the furnace is impure
lead which forms a third part of the vein
originally put in; when this is again fused it
gives black lead, having lost two-ninths in
bulk...The substance of white lead (tin) has
more dryness, whereas that of black lead is
entirely moist.

He continues(162):

Black lead which we use to make pipes and
sheets is excavated with considerable labour in
Spain and through the whole of the Gallic pro-
vinces, but in Britain it is found in the surface stratum of the earth in such abundance
that there is a law prohibting the production
of more than a certain amount.   The various
kinds of black lead have the following names -
Oviedo lead, **Capraria** lead, **Oleastrum** lead,
though there is no difference between them provided the slag has been carefully smelted away.

And again

There is also **molybdaena**(163) (which in another
place we have called **galena**);(164) it is a
mineral compound of silver and lead.   It   is
better the more golden its colour and the less
leaden: it is friable and of moderate weight.
When boiled with oil it acquires the colour of
liver.   It is also found adhering to furnaces

in which gold and silver are smelted; in this
case it is called metallic sulphide of lead.

Cinnabar (minium)

Some minerals were accidentally discovered:
"Cinnabar", explains Pliny(165) "was discovered in
the search for silver and a use was soon found for
the mineral." He closely follows the account of
cinnabar given by Theophrastos.(166) The description of the origin, the location and circumstances
of the discovery of the mineral is a direct
translation of the original text, slightly abridged.
This once more confirms Pliny´s direct use of many
of his sources, as Zehnacker(167) asserts (Pliny
disregards the evidence of Vitruvius(168) that
cinnabar was no longer exploited in the vicinity of
Ephesus in his time): "Theophrastos" writes
Pliny,(169)

states that cinnabar was discovered by an
Athenian named Callias, 90 years before the
archonship of Praxibulus at Athens. Callias
was hoping that gold could be extracted by
firing from the red sand found in silver mines;
and this was the origin of cinnabar although
cinnabar was being found even at that time in
Spain, but a hard and sandy kind and likewise
in the country of the Colchians on a certain
inaccessible rock from which the natives
dilodged it by shooting javelins, but that this
is cinnabar of an impure quality, whereas the
best is found in Cilbian territory beyond
Ephesus, where the sand is of the scarlet
colour of the **kermes** - insect; and that this is
ground up and then the powder is washed and the
sediment which sinks to the bottom is washed
again; and that there is a difference of skill,
some people producing cinnabar at the first
washing while with others this is rather weak
and the product of the second washing is the
best...

Juba(170) reports that cinnabar is also
produced in Carmania, and Timagenes says it is
found in Ethiopia as well, but from neither
place is it exported to us, and from hardly any
other place except from Spain, the most famous
cinnabar mine for the revenues of the Roman
nation being that of Sisapo (Almaden)(171) in
the Baetic region, no item being more carefully
safeguarded; it is not allowed to smelt and
refine the ore on the spot, but as much as

> 2000lbs. a year is delivered to Rome in the crude state under seal, and is purified at Rome, the price in selling it being fixed by law established at 70 sesterces a pound, to prevent it from going beyond limit.

Pliny(172) adds "There is in fact another kind of **minium** found in almost all silver-mines, and likewise lead-mines, which is made by smelting a stone which has the veins of metal running through it and not obtained from the stone, the round drops of which we have designated quicksilver (**argentum vivum**) - for that stone also if fired yields quicksilver - but from other stones found at the same time."

Masks were worn as a protection against the poisonous dust from cinnabar(173): "Persons polishing cinnabar in workshops tie on their face loose masks of bladder-skin, to prevent their inhaling dust in breathing which is very pernicious and nevertheless to allow them to see over the bladders."

Iron
The properties of magnetite aroused the interest of the Greeks(174) and Romans alike. Pliny(175) writes

> What is more strange than this stone (**magnetite**)...for iron is attracted by the magnet. This mineral is called by the Greeks another name, the ´iron stone´, and by some of them the ´stone of Herakles´. According to Nicander, it was called **magnes** from the name of its discoverer, Magnes, who found it on Mt. Ida... Sotacus (who wrote a work on stones in the third century BC) describes five varieties of magnetite:"an Ethiopian, another from Magnesia, which borders on Macedonia and is on the traveller´s right as he makes for Volos from Boebe; a third from Hymettus in Boetia; a fourth from the neighbourhood of Alexandria in the Troad; and a fifth from Magnesia in Asia Minor".

Some varieties do not exert attraction and the worst kind has been interpreted as a form of talc (the **magnetis lithos**) of Theophrastos.(176) Pliny(177) included **haematites magnes** "found in the sandy district of Ethiopia, known as Zmiris... which is blood-red in colour and when ground produces not only blood-red but also saffron-yellow powder". Eichholz(178) suggests that this is possibly red and brown haematite found together, possibly goethite, a species of brown haematite.   Pliny(179) continues "The test of an Ethiopian magnet is its ability to attract another magnet to itself... Also in Ethiopia and at no great distance is another mountain, (the ore from) which on the contrary repels and rejects all iron".   He refers, of course, to the attraction of unlike poles and the repelling of like poles, without understanding the the principles involved in this phenomenon.   Sotacus, the authority on whom Pliny directly relies, had been confused and this confusion is perpetutaed by him.

Pliny(180) also records varieties of **h(a)ematite**: "Among the oldest authorities Sotacus records five kinds of haematite, apart from magnetite": these are, Ethiopian, **andomas** (specular iron ore), Arabian (including **schistos**), limonite, **hepatite (kidney** ore) and **schistos**.

Among a number of other minerals recorded in the **Natural History** XXXV, sulphur, alum and asbestos are of particular interest.

Sulphur
Pliny(181) describes sulphur:
>Among other kinds of ´earth´ the one with the most remarkable properties is sulphur, which exercises great power over many other substances.   Sulphur occurs in the Aeolian islands between Sicily and Italy, which we have said are volcanic, but the most famous is the island of Melos.   It is there dug out of mine shafts and dressed with fire.   There are four kinds: ´live´ (**native**) sulphur (**vivum**), the Greek name for which means ´untouched by fire´ (**apyron**), which alone forms as a solid mass - for all the other sorts consist of liquid and are prepared by boiling in oil; live sulphur is dug up and translucent (**tralucet**) and of a green colour; it is the only one of all kinds which is employed by doctors.   The second kind is called ´clod sulphur´ (**glaeba**) and is commonly found in fullers´ workshops.   The third kind is only employed for one purpose, for smoking woollens from beneath, as it bestows whiteness and softness; this sort is called **egula**.   The fourth kind is especially used for making lamp-wicks.

In this passage Pliny says that ´natural´ (**vivum**)

sulphur(182), or rhombic, is found in volcanic areas. The gangue materials are removed by liquation, a process in which the sulphur is melted and run off from the accompanying impurities,(183) Oil is a solvent, as indeed are naphtha, pitch, benzine and similar hydrocarbons. Pliny is clearly unaware of the allotropic forms(184) of sulphur: ´clod sulphur´ is not likely to be what is now known as ´plastic´, since this is prepared by melting sulphur and suddenly chilling it. The use of sulphur dioxide (SO2) as a bleaching, or cleaning agent(185) accounts for its appearance in fullers´ workshops. With regard to the fourth category Pliny has misunderstood his source. Sulphur could obviously not be used for this purpose since the result would be much the same as lighing a ´fumite´ bomb. Perhaps the most interesting feature of this section is that the account has not been derived from Theophrastos.

Alum(186)

Not less important, or very different (from sulphur) is the use made of alum, by which is meant a salt exudation from the earth. There are several varieties of it. In Cyprus there is white alum and another sort of a darker colour, though the difference of colour is only slight; nevertheless the use made of them is very different, as the white and liquid kind is most useful for dyeing woollens a bright colour whereas the black kind is best for dark or sombre hues. (AlSO4). Black alum (that is, alum containing metals) is also used for cleaning gold. All alum is produced from water and slime... It occurs in Egypt, Armenia, Macedonia, Pontus, Africa and the islands of Sardinia, Melos, Lipari and Stromboli. The most highly valued is in Egypt and the next in Melos. The alum of Melos is also of two kinds, fluid (in solution) and dense (crystal). The test of the fluid kind is that it should be of a limpid milky consistency, free from grit when rubbed between the fingers and giving a slight glow of colour; this kind is called in Greek **phorimon**, in the sense of abundant. The other kind is the pale rough alum which may be stained by oak gall also and consequently this is called **paraphoron**, ´perverted´ or adulterated alum (the latter is possibly halotrichite, that is, hydrated iron sulphate with aluminium and ferrous sulphate). Liquid alum has astringent (as **stypteriodes ge**), hardening and corrosive properties... One kind of solid alum which is called in Greek **schistos**, ´splittable´, splits into a sort of filament of whiteish colour, owing to which some people have preferred to give it the Greek name, **trichitis**... There is another alum of a less active kind, called in Greek **strongyle**, ´round alum´.

Pliny seems to include aluminium sulphates, iron sulphate and potash-alum; also kaolinite and perhaps halotrichites.

Amiantus
"**Amiantus**(187) which looks like alum is quite indestructible by fire. It affords protection against all spells, especially those of the Magi."
The last book of the **Natural History**, XXXVII, is devoted to a discussion of gemstones. "Here" writes Pliny(188) "Nature´s grandeur is gathered together within the narrowest limits... hence very many people find that a single gemstone alone is enough to provide them with a supreme and perfect aesthetic experience of the wonders of Nature." The first major topic is the mineral used in the manufacture of ´myrrhine´(189) (**myrrhina, murrina, murrea** and similar adjectives), vases. The identifcation of this mineral well illustrates the necessity for the conflation of philology, archaeology and science in the study of Pliny´s text.

Fluor-spar
Modern editors of the **Natural History,** including H Zehnacker,(190) and D E Eichholz,(191) although generally accepting that the mineral was fluor-spar, still refer to the possibility that the vases may have been carved from agate, as suggested by Bauer(192): some editors likewise believe that Propertius(193) confused the two minerals. The evidence of Dionysius Periergetes(194) (**onychie litheia kai mourrine**) is cited in support of this belief. References, however, in the **Natural History**(195), the extant Crawford vases of either the first century BC or AD in the British Museum and evidence from the working of ´Blue John´(196), a variety of fluor-spar mined at Castleton in Derbyshire since Roman times, when there was a settlement at Anavio some three miles distant, all provide positive indications of the mineral described.

Pliny(197) follows the long standing theory about the formation of fluor-spar, **umorem sub terra putant calore densari,** that it is a liquid solidified underground by heat, the reference being to Aristotle´s dry exhalation, which he states is responsible for the hardening and colouration of the mineral. The nameless consul,(198) who gnawed the rim of his priceless cup, indirectly provided evidence of its low hardness. A material capable of being damaged by teeth (4+ relative to the Mohs scale of hardness) is clearly more likely to have been fluor-spar (Mohs 4) than agate (Mohs 7).(199) Pliny´s consul is the first person whose teeth enjoyed the benefit of flouride! Moreover, cups and ladles were generally described as brittle (the term used was **fragilitas.**(200)

Pliny(201) describes the vivid colours of the **vas myrrhina** as "purple and white disposed in undulating bands and usually separated by a third band in which the two colours being mixed, assume the tint of flame". Elsewhere he likens the colours of some types of glass to those of fluor-spar.

The significance of the hitherto incorrectly interpreted statement of Propertius(202): **murreaque in Parthis pocula cocta focis** is now clear. Romans employ two distinct technical terms to translate the process of heating, **calefacere** and **coquere:** the latter, however, implies a resultant chemical change. The cups were not fired but heated in front of fire (**in focis**, on the hearth) to modify their colour: this accords with the original view of Bailey.(203) Neither Pliny not Propertius had in mind Indian agate. The reason for the colours, as for example in the case of ´Blue John´ has been the subject of continuous research and speculation. The most recent study by Professor R N Haszeldine,(204) of the University of Manchester Institute of Science and Technology, suggests that it arose from the special geological history of Treak Cliff (Derbyshire), where radioactivity from ores of uranium type, thousands of years ago, displaced some of the calcium atoms from their normal positions in the lattice of the colourless calcium flouride crystals. The calcium atoms displaced in this way form colloidal groups of calcium atoms which aggregate preferentially in the defects, or irregularities of the lattice. The blue colour is formed by the scattering and by the absorption of light by the aggregates of colloidal atoms as light passes through the crystals of calcium flouride.

In practice heating a dark specimen of calcium flouride - often blue-black in colour - lightens its colour into a translucent amethyst purple with dark veins. The degree of heat, however, is critical, since too much will disturb the crystalline structure and result in the fluor-spar becoming opaque and off-white. The mineral is painstakingly mined by hand, since the shock of even controlled blasting would reduce ther fluor-spar to a coarse dust. Pliny observes that the brittleness of the cups inceases their value.

The nature of the mineral source explains the high value of cups and vases which were produced from single pieces of fluor-spar. Large pieces, or nodules, are rare, and veins normally vary from a quarter of an inch to a foot, with a depth of between 12 and 18 inches, the average being three and a half inches.

Myrrha itself does not refer to the actual mineral but to the resinous substances, or wax, used to impregnate the crystalline material before being worked. Pliny(205) mentions that ´myrrhine´ vases have a characteristic ´smell´, and further light is thrown on the observations of Martial,(206) who among a number of similar references, states: "If you drink your wine warm, Murrhine suits the burning Falernian and better flavour comes therefrom to the wine." This would seem to indicate an interaction between the vases and the wine. The wine would have gradually dissolved any residual traces of resin, producing perhaps something akin to the flavour of retsina.

Colour, therefore, the physical characteristics of the mineral, as described by Pliny, including its relatively low degree of hardness, the nature of the vases and the effect of heat, all contribute to the inevitable conclusion that ´myrrhine´ ware was made of fluor-spar.

Adamas

The diamond (**adamas**) was known to the Romans certainly from the principate of Augustus,(207) but the term **adamas** in the ancient world was used of virtually any hard material, whether mineral or metal. Pliny(208) enumerates some of the sub-stances so described:

> The most highly valued of human possessions, let alone gemstones, is the **adamas**, which for a long time was known only to kings, and to very few of them. **Adamas** was the name given to the

'knot of gold' (**auri nodus**) found occasionally
in mines in association with gold and, so it
seemed, formed only in gold. Our ancient
authorities thought that it was found only in
the mines of Ethiopia between the temple of
Mercury and the island of Meroe (northern
Sudan), and stated that the specimens
discovered were no larger than a cucumber seed
(**cucumis semine**) and not unike one in colour.
Now for the first time, as many as six kinds of
adamas are recognised. There is the Indian
which is not formed in gold and has a certain
affinity with rock-crystal... Similar to the
Indian, only smaller, is the Arabian, which is
moreover formed under similar conditions. The
rest have a silvery pallor and are liable to be
formed only in the midst of the finest gold...
One of these stones is called in Greek **cenchros**
or millet seed (**milii magnitudine**), which it
resembles in size. A second is known as the
Macedonian and is found in the gold mines of
Philippi. This is equal in size to a cucumber
seed. Next comes the so-called Cyprian, which
is found in Cyprus and tends towards the colour
of copper,(209) but has potent medical
properties... After this there is the
siderites, or iron stone, which shines like
iron and exceeds the rest in weight, but has
many different properties. For it can not
only be broken by hammering but can also be
pierced by another **adamas**. This can happen
also to the Cyprian kind, and, in a word, these
stones being untrue to their kind possess only
the prestige of the name they bear.

Diamond (210)

Only the Indian and, possibly, the Arabian are true
diamonds. The first certain literary reference to
diamonds occurs in Manilius,(211) who writes: "Thus
the diamond, a stone no bigger than a dot, is more
precious than gold." Roman trade with the East
flourished under the empire. The earliest diamonds
were brought to Rome from India where they were
found in "diamantiferous sandstone...which belongs
to the oldest division of sedimentary formations of
the country which usually rest directly upon the
still older crystalline rocks such as granite,
gneiss, mica and other schists (hornblende-,
chlorite-, and talc-). The diamonds are found in
any earthy bed containing abundant pebbles. The
diamond river mentioned by Ptolemy may have been the

Mahandi (SW of Calcutta) in the Hyderabad area: the
river also yields an appreciable amount of alluvial
gold."(212)   Pliny(213)   lists   the   physical
properties of diamonds: "All these can be tested
upon the anvil and they repel blows so that an iron
hammer head may split in two and even the anvil be
unseated.   Indeed   the   hardness of the diamond is
not able to be described and so too the property
whereby it conquers fire and never becomes heated."
Pliny confuses the hardness of diamond with its
brittleness.   The   term **adamas** is, in fact, sing-
ularly inappropriate and inaccurate when its extreme
frangibility, or brittleness, is considered.   A
diamond is easily fractured - a very moderate blow
from a hammer sufficing   for the purpose: its
perfect cleavage places it among the most brittle of
minerals.(214)   Diamonds are, however, unaffected
by heat unless it reaches about 700° C.(215)
Pliny,(216)   however,   contradicts   himself   by
observing,   elsewhere,   that   broken   diamonds
disintregate into splinters so small as to be
scarcely visible.   These are much sought after by
engravers of gems   and are inserted by them into
iron tools because they make hollows in the hardest
materials without difficulty.   Diamond is 10 on the
Mohs scale(217) of hardness, ninety times harder
than corundum (Mohs 9).   The term ´Arabian´ may be
a trade name, or refer to diamonds of an inferior
quality originating in India but re-exported by way
of Arabia.
    The identification of the other varieties of
adamas listed by Pliny presents a number   of
difficulties.   It has been suggested that **cenchros**
was the name for a small diamond,(218) possibly of
the type in Manilius´ reference.   **Adamas** as ´knot
of gold (**auri nodus**) and the Macedonian **adamas**
appear to be the same substance, although the latter
has been equated with rock-crystal.(219)   Another
possibility has been considered, namely that the
Macedonian   substance is platinum, a metal mainly
known from Egyptian sources.   Platinum   group
elements (PGE)(220) are normally associated with
alluvial gold (placer deposits) although rarely with
deep mined, reef gold.   Support for the presence of
traces of platinum in gold, however, is afforded by
staters of Philip II in which PGE occur as tabular
inclusions some 35 microns in length,(221)   On
balance, however, the evidence from literary and
archaeological   sources   is   against   such   an
identification.   Finally, Pliny´s description of
the Macedonian **adamas** is less conclusive than has

been suggested since platinum is greyish in colour, nor does it resemble a cucumber seed. The Cyprian **adamas** may have been **analcime**, that is, hydrous sodium aluminium silicate.(222) **Siderites** presents no difficulty: this is simply iron pyrites.

Beryl
The mineral beryl affords a good example of the difference in attitude to classification.(223) The mineralogist includes the deep green, bluish green, greenish blue and yellow specimens all in the same species to which he gives the name **beryl**, since they all agree in chemical composition ($3Be.Al2O3.SiO2$), being a silicate of the metals beryllium and aluminium, and crystalline in form, differing only in colour. The gemmologist however refers to the deep green variety as emerald, to the greenish blue and bluish green varieties as aquamarine and **only** to the yellow stone as **beryl**. Pliny(224) correctly observes: "Many people consider the nature of beryls to be similar to, if not identical with, that of emeralds."

Tourmaline
Pliny, following Theophrastos, describes two sub-stances, other than magnetite, as having the power to attract other materials. One is amber,(225) the other is the mineral tourmaline.(226)

> To the same class of fiery stones belongs the **lychnis** so called from the kindling of lamps, because at that time it is exceptionally beautiful...
> I found that there are other varieties as well, one of which has a purple, and the other a scarlet sheen (these are violet-red and rose-red tourmaline respectively). These when heated in the sun or by being rubbed between the fingers are said to attract straws and papyrus fibres. It is said that the same power is exerted by the Carthaginian stone, although it is far less valuable than those previously mentioned.

Tourmaline(227) is a boro-silicate of very com-plicated composition in which, in contrast to the chemical composition, the crystal forms of all varieties are in close ageement.
Pliny was clearly aware of the pyro-electric properties ot tourmaline. Friction is adequate to produce static electricity but a change of

temperature allows the mineral to be electrified more readily. This phenomenon is intimately related to the hemimorphic, or polar development of the crystals. On being warmed, one end becomes charged with positive electricity, and the other end acquires a negative charge. On cooling the charges are reversed.(228)

Tourmaline also has piezo-electrical properties: when a plate, cut perpendicular to the principal axis is subjected to variations in pressure, it develops positive and negative charges on the two surfaces. This additional property was unknown to Pliny. Topaz,(229) like tourmaline, is remarkable for the strength of the pyro-electrical charge it may acquire and for the length of time it can hold this - about 30 hours. Similarly it has piezo-electric properties and pressure of the fingers exercised in the direction of the length of the prismatic crystal is sufficient to create a charge. Pliny, however, did not identify these properties in topaz or a number of other precious stones which may likewise acquire feeble, fast disappearing charges. The presence or absence of this characteristic may be an important factor in distinguishing gems of like appearance, for example red tourmaline from ruby and greenish blue topaz from aquamarine. Neither Pliny, nor other ancient writers on minerals followed up this discovery, or applied it to classification.

Conclusion

In succeeding centuries works on minerals continued to appear sporadically. Avicanna (980-1037 AD), the Arabian translator of Aristotle, grouped minerals as "Stones, sulphur minerals, metals and salts." In 1262 Albertus Magnus wrote his **Natural History,** five books of which were devoted to minerals, but only those which bestowed supernatural powers on their owners were considered valuable. Other works followed until Georgius Agricola, who was born in the mining region of Erzgebirge in 1491 (d.1555), and who had become a keen observer and careful recorder of minerals, formulated the first reasonable theory of ore genesis.

Pliny the Younger,(230) in a letter to the historian Tacitus, writes of his uncle "The fortunate man...is he to whom the gods have granted power either to do something which is worth recording, or to write what is worth reading, and most fortunate of all is the man who can do both.

Such a man was my uncle, as his own books and yours
will prove."

The modern science of mineralogy was not
established until the early part of the nineteenth
century, when it was first recognised that chemical
composition and crystalline form were of the first
importance and that external characteristics were
often accidental.

In books XXXIII to XXXVII of his **Natural History,**
however, Pliny´s discussion of minerals, earths,
stones and gemstones makes an important contribution
to our knowledge of Roman achievement in a major
area of Earth Sciences. Pliny advanced the subject
beyond the discoveries of his predecessors and his
work on minerals had a long-lasting influence.

NOTES

1. Para. 13
2. Herodotus, I,1 applies the term **historie**
to his writings. It is subsequently used by the
preSocratic philosophers in the sense of ´knowledge
resulting from enquiry´.
3. I,53ff. Cf. also Virgil, **Georgics** II, 490
4. **Preface,** para 14
5. See R Syme,"Pliny the procurator",**Harvard
Studies in Classical Phology,**73, 1969,pp.201-236
6. J M Blazquez, **Explotacions mineras en Hispania durante la republica y el alto imperio romano.
Problemas enconomicos, sociales, y tecnicos,** Madrid,
1969; P R Lewis and G D B Jones, "Roman gold-mining
in north-west Spain", JRS, 60, 1970, pp. 169-185,
with a translation of Pliny **NH** XXXIII. 67-78; R F
Jones and D G Bird, "Roman gold-mining in north-west
Spain. II, workings on the Rio Duerna", Ibid., 62,
1972, pp. 59-74; D G Bird, "The Roman gold mines of
north west Spain", **BJ,** 172, 1972, pp.36-64
7. "either rustic terms, or foreign, indeed
even barbarian words which have to be introduced
with an apology": **Preface,** para. 13. See also J
Healy, "Problems in mineralogy and metallurgy in
Pliny the Elder´s Natural History", **Tecnologia,
economia e societa nel mondo romano,** Atti del
Convegno di Como, 27-29 September 1979, p.184
8. **NH** XXXVII.23-29, esp. 26
9. Ibid., 56
10. Ibid., 76-79
11. Ibid., 136f.
12. **Traité de crystallographie,** 1822
13. **Traité des caractères physiques des**
pierres précieuses, 1817

14.   **De Lapidibus** para. 28 and pp. 116ff.
15.   **NH** XXXVII.10
16.   W Gilbert, **De Magnete, Magneticisque Corporibus, et de Magno Magnete tellure, Physiologica Nova**, London, Peter Short, 1600.
17.   J F Healy, **Mining and Metallurgy in the Greek and Roman World**, London, Thames and Hudson, 1978, p, 15
18.   Ibid., p.16ff
19.   D Whittern and J Brooks, **A Dictionary of Geology**, London, 1974, pp. 353ff.
20.   G K Guthrie,**A History of Greek Philosophy**, Cambridge, 1962, vol.I, pp. 39ff.
21.   Ibid., pp.58ff.
22.   Ibid., pp. 122-265
23.   Ibid., pp. 1-80
24.   **Timaeus** 60b 7 - c 6.   Cf. also R Halleux, "Le problème des métaux dans la science antique", **Bibliothèque de la faculté de philosophie et lettres** de l´université de Lièges, CCIX, 1974, pp. 65ff.
25.   **Meteorologica** III.378a-b
26.   See generally, D E Eichholz in his ed. of Theophrastos, **De Lapidibus**, Oxford, 1965, pp. 38ff., esp. 39, and previously, "Aristotle´s theory of the formation of metals and minerals", **QC** 43, 1949, pp. 140ff.
27.   **Meteorologica** III.383b
28.   Theophrastos, **De Lapidibus**, ed Eichholz, p. 50
29.   **Meteorologica** III.383b
30.   Eichholz (n.28 above) p. 47
31.   Ibid., and E R Caley and J F Richards, **Theophrastus on Stones**, Columbus, Ohio, 1956
32.   **De Lapidibus** paras. 1 and 3, Caley and Richards, op.cit., p. 45.   See also commentary, pp. 63ff.   Cf Aristotle, **Meteorologica** IV. 6-12
33.   Ibid.
34.   **NH** XXXIII.113-4
35.   **De Lapidibus**, ed. Eichholz, paras. 58-9
36.   L Edelstein and I Kidd, **Posidonius. I The Fragments**, 1972
37.   Diogenes Laertius, 58f. mentions a lost work of Straton (340/330 - c. 270 BC) on **Mining Tools and Machines**.   Cf. also H B Gottschalk, "Strato of Lampsacus: some texts", **Proceedings of the Leeds Philosophical and Literary Society,** Literary and Historical Section, 11, 1965, n.6, pp. 95 -182, and Halleux, **Metaux**, pp. 115ff.
38.   **NH** XXXVI.161; XXXVII.21, 23, 26-8, 48
39.   D Eichholz, Pliny, **NH** vol.X (**LCL**), intro., xv and ´A D Nock, **JRS** XLIX, 1959, p. 14

40. **NH** XXXVII.23 - 26.   This theory may also owe something to Posidonius.

41. **Naturales Quaestiones** III.25, 10

42. II.52, 1-2:   In these countries are generated not only animals which differ from one another in form because of the helpful influence and strength of the sun, but also outcroppings of every kind of precious stone which are unusual in colour and and resplendent in brilliancy.   For the rock crystals, so we are informed, are composed of pure water which has been hardened, not by the action of cold, but by the influence of a divine fire, and for this reason they are never subject to corruption and take on many hues when they are breathed upon.

43. **NH** XXXVII.21 and M Bauer, **Precious Stones**, New York, 1968, vol. 2, p.474; see also R Halleux, "Fécondité des mines et sexualité des pierres dans l´antiquité greco-romaine", **Revue belge de Philologie et d´histoire**, 42. 1970, pp. 16-25. K C Bailey, **The Elder Pliny´s Chapters on Chemical Subjects**, London, 1923, vol. 2 p. 253, writes "It is not surprising that the growth of crystals should have been considered in early times to have had some kinship with organic growth".

44. **NH** XXXVII.21

45. **NH** XXXVII.48

46. E Warmington, **The Commerce between the Roman Empire and India**, Cambridge, 1928, pp. 235ff. (mineral products).

47. **NH** XXXVI.161

48. **NH** XXXV.184

49. **NH** XXXVI.125

50. **NH** XXXIV.164

51. V.2,6.   He also mentions marble quarries on Paros and salt mines in India.

52. Bauer, **Precious Stones**, I, 143

53. **Letters** III.5,7ff.

54. **NH** XXXVII.169.   Cf also XXXVI.144

55. **NH** XXXVII.186

56. **NH** XXXVII.23

57. Ibid., 57

58. Ibid., 146

59. **NH** XXXVI.139

60. **NH** XXXIII.95

61. **NH** XXXIV.159

62. **NH** XXXVI.168.   Cf Livy, XLI.27,5

63. **NH** XXXVIII.103

64. XVII.3.11

65. **NH** XXXVI.14

66. Ibid.

67. Bailey, n. 43 above.

68. Dioscorides, V.127
69. (XXXIII) p. 182
70. NH XXXVII.148
71. Eichholz, NH vol.X (LCL), p. 118, note b.
72. Cf. XXXIII.84
73. NH XXXVII.55ff.
74. Ibid., 56
75. Ibid., 57
76. Ibid., 58
77. J Ogden, "Platinum group metals inclusions in ancient gold artifacts", Journal of Historical Metallurgy Society, 11, n.2, 1977, pp.53ff.
78. NH XXXVI.146
79. Theophrastos, De Lapidibus, para 26, and NH XXXIII.4 and 86
80. Ibid.
81. Whitten and Brooks, n. 19 above, p.82, appendix.
82. Healy, Mining and Metallurgy, p. 74
83. NH XXXVI.128
84. Bauer, Precious Stones, vol.I, p. 55
85. NH XXXVII.131
86. Bauer, Precious Stones, vol.I, pp. 11-68, gives a detailed account. See also C S Hurlbut, Dana's Manual of Mineralogy, pp. 126ff.
87. NH, XXXVI.131
88. Ibid., pp. 160-162
89. Cf. XXXVII.23
90. NH XXXVI.132. Cf. Theophrastos, De Lapidibus para.6
91. NH XXXVI.132 and Theophrastos, De Lapidibus, para 7
92. NH XXXIV.173
93. NH XXXV.192
94. Healy, Mining and Metallurgy, p. 28
95. NH XXXVI.141-2
96. NH XXXVII.72
97. NH XXXIII.59 and 61
98. Although Herodotus (I.50,2) had already in the fifth century BC observed the relationship between mass and volume.
99. NH XXXIV.157
100. G W C Kaye and T H Laby, Tables of Physical and Chemical Constants, 1966, pp. 118ff.
101. Whitten and Brooks, n. 19 above, appendix (quartz)
102. NH XXXIII.61
103. Timaeus, 59b
104. NH XXXVI.127. Cf.XXXIV.147
105. NH XXXVII.53. So Theophrastos, De Lapidibus, para. 28.

143

106. **NH** XXXVII.103
107. Whitten and Brooks, n.19 above, pp. 431-2
108. Healy, **Mining and Metallurgy**, p. 29
109. **NH** XXXV.192
110. **NH** XXXVI.147
111. This is part of Pliny´s technical vocabular
112. **NH** XXXIII.126
113. **NH** XXXVI.133
114. Eichholz, **NH**, vol. X (**LC**), p. 107, note e.
115. Whitten and Brooks, n. 19 above, pp. 192
and 340 respectively.
116. I am indebted to Dr R C A Röttlander,
Tübingen, for advice on this technical distinction.
117. **NH** XXXVII.121, 126 and elsewhere.
118. **NH** XXXI.84 and XXXVIII.184
119. **NH** XXXVII.56, 129, 158
120. **NH** XXXVI.163
121. **NH** XXI.53
122. **NH** XXXVII.66
123. Ibid., 69-70
124. Ibid., 115
125. Ibid., 105
126. Ibid., 79
127. Ibid., 21
128. Ibid., 106
129. Ibid.
130. **NH** XXXVII.26
131. Ibid., 55ff.
132. Bauer, **Precious Stones**, vol.I, p. 119
133. **NH** XXXVII.76
134. Ibid., 79
135. Ibid., 136-7
136. Ibid., 138
137. Ibid., 196
138. For Pliny´s career, see Healy, **Inter-
disciplinary Science Reviews**, vol.6, n.2, pp. 166-7
139. **NH** XXXIII.62
140. Ibid., 66
141. Cf. Herodotus, III.102-5
142. **Preface**, para. 13
143. A H M Jones, **A History of Rome through the
Fifth Century, I. The Empire**, p. 300ff.
144. **NH** XXXIII.80
145. III.14,1ff. See Healy, "Mining and processing of gold ores in the ancient world", **Journals
of Metals (AIME)** vol. 31, 1979, pp. 11-16
146. Ibid. and J H Notton, "Ancient Egyptian
gold refining", **Gold Bulletin**, vol. 7, April 1974,
pp. 50-56
147. III.2,8
148. **NH** XXXIII.84

149. Originally published by by C Leemans,
**Papyri Graeci Musei Antiquarii Publici Lugduni
Batavi**, Leyden, 1885, pp. 209-249
150. XXXIV.121
151. V.101
152. **NH** XXXV.186
153. note 43, above, vol.II, p. 233
154. **NH XXXIII**, p.182
155. **LS(9) (Schistos)**
156. **NH** XXX.126
157. **De Lapidibus** para. 46-7
158. Healy, **Mining and Metallurgy**, p. 205 and K
Hradecky, **Die Strichprobe der Edelmetalle**, Vienna,
1930
159. **NH** XXXIII.95
160. Ibid., 106
161. **NH** XXXIV.159. See generally, R Halleux,
"Les deux métallurgies de plomb-argentifère dans
l'histoire naturelle de Pline", **Revue de philologie**,
44, 1975, pp.72-88. Pliny utilised two lost
sources: (i) Cornelius Bocchus (**NH** XXXIII.96 and
XXXIV.159) and (ii) Sextius Niger (XXXIII.106-8 and
XXXIV.173) Cf. also Polybius XXXIV.9,10
162. Ibid., 154
163. Ibid., 173
164. **NH** XXXIV.159
165. **NH** XXXIII.111-124, esp. paras. 112-114
166. **De Lapidibus**, paras.58-9
167. **NH** XXXIII, pp. 200ff.
168. **De Architectura** VII.8.1
169. **NH** XXXIII.113
170. Ibid., 118
171. Ibid.
172. **NH** XXXIII.99
173. Ibid., 122
174. Theophrastos, **De Lapidibus**, para. 29
175. **NH** XXXVI.126-7
176. **De Lapidibus**, para. 41
177. **NH** XXXVI.129
178. **NH** vol. X (**LCL**), 103, note e
179. **NH** XXXVI.129
180. NH XXXVI.146
181. **NH** XXXV.174-177
182. Ibid., 175. See further J W Mellor, **A Comprehensive Treatise on Inorganic and Theoretical
Chemistry**, London, 1930, vol. 10, pp. 1-2 and 14.
The amount of sulphur in "workable" ore varies from
eight percent up to about 25.
183. Ibid., pp. 14-15
184. Ibid., pp. 23ff.
185. Ibid., p. 243

186. **NH** XXXV.183-4. See Bailey, note 43 above, vol.2, p.233
187. **NH** XXXVI.139
188. **NH** XXXVII.1
189. Ibid., 18-22
190. (XXXIII.5), pp. 121-2
191. Vol.X (**LCL**) (XXXVII.21), p.178, note a
192. **Precious Stones**, vol. 2. pp. 530-1
193. IV.5.26
194. **Periplous Maris Erythraei**, 49
195. **NH** XXXVII.18-22
196. Bauer, **Precious Stones**, vol.2, p.530
197. **NH** XXXVII.21
198. Ibid.,18
199. Whitten and Brooks, note 19 above, p.221 and below p. 223
200. **NH** XXXIII.5
201. **NH** XXXVII.21-2.   See also Bauer, **Precious Stones**, vol. 2, p. 530
202. IV.5.26
203. Note 43, above, vol. 1
204. A E Ollerenshaw and R J and D Harrison, **The History of Blue John Stone**, p. 24
205. **NH** XXXVII.22
206. **Epigrams**, XIV.113
207. Warmington, note 46 above, pp. 235ff. (mineral products)
208. **NH** XXXVII.55
209. Ibid., 58
210. Ibid., 56
211. **Astronomicon** IV.926
212. Bauer, **Precious Stones**, vol. 1, pp. 141, 149ff.
213. **NH** XXXVII.57
214. Bauer, **Precious Stones**, vol. 1, pp. 128ff.
215. Ibid., 115
216. **NH** XXXVII.60
217. Bauer, **Precious Stones**, vol.1, p. 33
218. Eichholz, **NH** vol. X (**LCL**), p. 208
219. Ibid.
220. J Ogden, note 77 above.
221. Ibid., p. 56, fig.2
222. Eichholz **NH** vol. X (**LCL**), p. 209
223. Bauer, **Precious Stones**, vol.1, p. 107
224. **NH** XXXVII.76
225. Ibid., 53
226. Ibid., 103
227. Bauer, **Precious Stones**, vol. 2, pp. 363ff.
228. Bauer, **Precious Stones**, vol. 1, p. 67
229. Ibid.
230. **Letters**, VI.16.3

Chapter Nine

CHEMICAL TESTS IN PLINY

F Greenaway

Pliny´s treatment of subjects we would now distinguish as chemical ranges over a wide field:
technical, pharmaceutical, metallurgical and much
else. So wide is this scatter of chemical passages
that it would be quite wrong to try to elaborate
some supposed Plinian view of chemical Nature. It
is impossible to pull them together by a common
thread of theory. Fifty years ago K C Bailey (1)
made a survey of the chemical passages which remains
the best guide to them, and there is not much one
can add to his comments or individual statements in
the text. But, all the same, a moderm attitude to
the history of science invites us to go further than
Bailey´s bare recital of glosses on Pliny´s facts.
We can seek to build on it some better idea than was
possible even so recently as Bailey´s day, of the
recognition by Pliny and his contemporaries of an
area of the natural world which had a particular
identity. We call it chemical (2) but no equivalent word would have been known to Pliny.
However, some such identity seems to have been
recognised by the alchemists,(3) who came later than
Pliny, in their theoretical view of the transformation of matter. Their quasi-chemical picture
of a soul-related physical world(4) was different
from that of the philosophers who preceded them and
different again from that of the technical chemists
who were always active. Nothing quite like any of
these systems emerges from Pliny, yet one is
conscious of a degree of consistency in his
approach.(5) I shall try to suggest one source of
consistency.
    To make sense of the chemical part of Pliny´s
compendium it seems wise to select one feature.
Let us consider decision-making. In any study of a
seemingly unorganised mass of information it is

147

often a good idea to pick out factors which were relevant to the making of decisions, a criterion which is particularly valuable when one is dealing with matters that have an economic component. So we may choose to examine Pliny´s work with an eye to the choice or selection of material substances entering into commerce, or for everyday use. It would be flying too high to think of this as fore-runner of analytical chemistry, but it is difficult to think of any continued civilised life using material substances which did not use many kinds of discriminatory tests. Reading Pliny with this in mind, we become aware that he draws our attention repeatedly to decision-making criteria which must have permeated the whole of the economy of his world. A modern mind would at once think in terms of analytical chemistry, one of the nerve fibres of an industrialised society.(6) We find nothing in Pliny that satisfies any modern definition of ´analytical chemistry´. To ´analyse´ something meant at first (and even in chemistry of the post-renaissance scientific epoch) to separate its parts. In modern terminology it means to characterise and to estimate them. Before the late 18th century learned men generally supposed they knew what substances they were handling. It is only after the maturing of scientific activity that chemists addressed themselves systematically to examining substances of unknown composition in the confident belief that that they could determine their composition. It will help us to examine Pliny if we bear in mind that, even in these days of elaborate chemical and chemico-physical procedures, there are really only three ways in which a chemical species may be detected and estimated(7):

1.   by isolation of the species itself;

2.   by inference from a characteristic chem-
     ical reaction of the species;

3.   by inference from a characteristic phys-
     ical property of the species.

I must not try to force what we find in Pliny too firmly into these moulds, but they will be helpful as a background.
     Pliny is not isolated. There is evidence in his times of a well developed practice of arts which were one day to underly technical chemistry. The well known Leyden and Stockholm (8,9) papyri (which are

evidently part of one document) are of a later period than Pliny, but very likely gathered together knowledge accumulated over several centuries. They are useful in authenticating many of the quasi-chemical episodes in Pliny, by showing from a different point of view that he was describing activities familiar to experienced practitioners. There is difficulty here: although we are quite ready to recognise sharp historical changes in, say, political matters, like the consequences of the death of a ruler, or intellectual matters like the preaching of an innovative philosopher, we are not able so easily to recognise sharp technical changes in the ancient world.(10) For this reason, we can, for the time being, suggest that Pliny would have been familiar with the kind of technique described in the Leyden/Stockholm papyrus. So let us suppose that this repertoire was in general use in Pliny´s day. Then, judging by the literary context, we can suggest that Pliny´s knowledge was wide compared with that of the practitioners who compiled the papyrus, but not necessarily more critical. He does not seem to have used sources of a higher standard of discrimination than his own, but the papyrus practitioners were weaker in one respect. Pliny is often concerned with making certain one has the right material to start with. The papyrus recipes (and it is really only a recipe book) are clearly based on the principle that if you do the right things to the right materials you will get the right results. However, there is never any instruction about making sure one has got the right material to start with. There are some tests for the finished product, some tests for ascertaining whether one has reached the end of a process, or for deciding when to go on from one stage in a process to the next.

Some tests in the Leyden/Stockholm papyrus will give us a rough idea of method and attitudes. There is a brief reference to "unfalsified proof silver". This suggests a standard of some sort, but it may well have been merely silver subjected to prolonged heating and therefore considered to be as pure as possible. An imitation of silver will "be of the first quality, except that artisans can notice something about it because it is formed by the procedure mentioned" (that is, a procedure for imitation). Evidently judgememt by eye was the most important test, but it was associated with the history of the object judged. The same applies to other imitation procedures and to the excellence of

dyeing.   In Pliny we find many more tests of one
kind or another, many of which are intended to
identify a material as correct for its intended use.
Following is a list of some types of test and the
number of times they appear more or less clearly
mentioned.

1.	Smell	1
2.	Taste	5
3.	Feel	3
4.	Colour	10
5.	Melting point	2
6.	Effect of heat	5
7.	Density (estimated)	1
8.	Density (measured)	2
9.	Chemical change	6
10.	Process test	1
11.	Touchstone	3

One has to use some such phrase as "clearly
mentioned" because the language does not always
enable one to identify the actions described, so
these figures are very rough.   (There is   in
addition one test which falls into no modern
category, namely the use of a beaker made of
electrum for the detection of poisons.) This list
shows that the criterion most often used to indicate
the nature or quality of a substance is just its
appearance.   Chemical qualities of a rough and
ready nature come next, then heat tests.   Density
is only very roughly used, but we should not dismiss
this lightly.   Artisans and craftsmen can become
very adept at estimating dimensions for which the
unskilled person would need a measure of some sort.
So Pliny may have read reports of, or seen tests
which look rough and ready now, but were reliable in
craft or commerce.   I do not want to try to dignify
them as scientific or even to suggest that there was
any orderly connection between craft judgements, but
the foundations of an orderly system seems to have
been there.

Tests described
Let us now comment on some of the tests described.

Flos salis (XXXI.91).   Immediately we come up
against a difficulty.   **Flos salis** was an ingredient
of unguents, the type of cosmetic most widely used
in the ancient and classical world.   We are still
uncertain what it was, but it seems to have been a
fatty or oily substance found in association with

some springs or salt deposits.   It was probably red
in   colour,   since the usual adulterant was ochre or
powdered potsherds.   The test for adulteration was
to wash with water, which dispersed impurities
leaving the desired **flos salis**, which was soluble
only in oil.   This test is much the same as a
purification process, which is true of many other
kinds of test or analysis even today, as we observed
in the first category of methods of analysis in the
opening remarks of this paper.
        There are other names(11) which we cannot
associate certainly with any chemical or geological
species.

Aphronitrum (XXXI.113) is one.   It may be the same
as **nitrum,** which could be soda, soda with salt, or
perhaps the nitre (potassium nitrate) of later
times.   Pliny states that he is quoting someone
else and there is no obvious clue.   However, the
quality of each substance is judged by its being
friable, of low density and possessing a purple
colour.   (Even "purple" may need to be qualified:
colours and colour changes were incapable of
definition in the classical languages in any way we
can easily relate to modern   colour-terms   and
changes(12).)   **Nitrum** is also said to be
characterised by lightness, sponginess and porosity.
Bailey discusses this at length because of the many
references to **nitrum** in classical literature which
sometimes make it appear to be soda, sometimes
potassium carbonate, sometimes even potassium
nitrate (as with **aphronitrum**).   It turns up in
accounts of unguents and detergents, of embalming
and glass making.   One remark of Pliny is
puzzling: he says that nitrum does not crackle in
the fire.   Potassium nitrate would certainly do so,
but some other salts would produce a crackling noise
by decrepitation.   It seems to me that Pliny is
badly reporting someone´s distinction between a
nitrate **nitrum** and some other kind.   (There is a
tempting speculative explanation: if we suppose that
the cave referred to as the source was in fact some
excavation or structure used as a human habitation
or place for livestock.   It has been established
that in the middle ages and later potassium nitrate
was found in, and collected from, country dwellings
where walls and mounds were impregnated with animal
and human urine, which after decomposition, produced
an efflorescence of potassium nitrate.   But this is
an unsupported guess.)(13)
        By   contrast   with   the   ambiguities   in   testing

nitrum we can point to one test which is not only unambiguous but has kept its form and meaning for centuries. This is the fire test for gold, varying from the simple form of a test literally in the fire, to sophisticated modern assaying. I have said something elswhere(14) about the historical continuity of the tradition of assaying, so I need only say here that Pliny is well aware that "alone of all substances it loses nothing on heating and survives even conflagrations and the funeral pyre". In fact, he asserts that the oftener it is heated the better it becomes. This sounds silly to us who are accustomed to the idea of a pure substance of 100 percent purity as a limit at which successive purification ceases. However, this idea of a pure substance is comparatively modern, and cannot be clearly identified until the years of the chemical revolution associated with the names of Lavoisier and Dalton. Even as late as Newton(15) the idea of endless purification is to be found, as for example, in the Gold Trial of 1710 following the Scottish coinage after the Act of Union - it was found that the trial plate was purer than any coin, but Newton held that "gold may be refined so high as to be about half a carat finer than 24 carats". There does not seem to have been any idea of an absolute purity independent of further treatment.

It is all too easy in translating technical matter into modern English even from the English of previous centuries, let alone from Latin, to use ideas which could have had no currency at the time. An instance of some importance occurs in this same chapter (XXXIII.60). Bailey has translated **Primum autem bonitatis argumentum quam difficilis accendi** as "but the best proof of purity is a high melting point". This is much too sophisticated an idea for the period, or indeed until modern times. A better translation would be "difficulty of melting" which is as much as the original will bear.

It is no more definite a property than the resistance of gold to corrosion which is known to be an outstanding property but does not constitute a test. This is, as it were, a surface property and we find one test described which relates to a surface condition of importance. Gilding(16) was widespread as a decorative art and it was clearly desirable that the gold amalgam used should take firmly on the surface to be gilded. Pliny describes the preparation of the copper by heating and quenching with vinegar. The surface is then examined to see if it is bright enough to respond to gold amalgam.

Electrum

We can mention here the alloy known as **electrum,** which consisted of silver and gold. Pliny says (XXXIII.80) that "all gold contains silver in varying amounts"; and says that when the proportion reaches one fifth silver it is known as **electrum.** An alloy of greater silver content is held to possess sufficient strength to be workable. At this point one comes up against that mixture of sound reporting and gossip mongering which makes the study of Pliny so exasperating. We read that a cup made of naturally occurring **electrum** is held to be a detector of poisons. When they are present, rainbow tints form in the cup, and a crackling noise like fire is heard and "so twin proofs are given". (XXXIII.80). It seems that Pliny was reading **electrum** as a silver/gold alloy when **electrum** was also in use to mean amber. The crackling associated with the electrostatic properties of amber may have crept in there and offer Pliny the chance of making a vivid piece of reporting devoid of any real technical basis.

He can be excused, however, for tripping over nomenclature. This was a problem for the learned from the earliest times, and in the field of chemistry, Crosland(17) gives many examples of doubts and confusion facing early workers and the present day historian.

Stimmi

It is not unusual to find what we recognise as different substances being treated as variants of the same substance, and for different kinds of reason. Let us take two examples.(18) One is **stimmi** (also known under different names which I shall omit for clarity). This is said to be found in silver mines and to exist in a male and female form. The male is course, rough and less dense. The female has a shining, smooth surface and cleaves lengthwise. These forms were antimony sulphide and metallic antimony. It is not at once apparent why these two distinct substances should be treated as one, but the subsequent description of the making of the **stimmi,** used as a cosmetic or medicament, could apply equally to either. The one was roasted with a paste of which cow-dung was a major constituent and the antimony (or the sulphide) would be converted to the oxide. In either case some free antimony would be deposited. The antimony oxide

would be washed out. It has the property of con-
tracting eyelids, producing the apparent enlargement
of the eyes which was considered beautiful (and so
familiar to us in Roman portrait paintings.

Minium
Another type of confusion was between substances of
like appearance usable for a single purpose. Here
we see several examples in pigments. **Minium** is a
long-lived term for a number of substances. It
could be used for cinnabar (HgS) the best material
for use as a pigment, or for the inferior red lead.
The distinction between the several kinds of **minium**
was of the simplest nature, apparently only the
weight of a standard package. One can hardly call
this a density measurement , but one can infer that
the trade was regular enough for so simple a
discrimination to be valuable. A more technical
test for adulteration (as distinct from differentiation) was to heat a specimen of **minium** on a
plate (he specified a gold plate, but this is
irrelevant). This heating darkens the adulterated
material. Pliny´s chemistry is inadequate here.
Both mercuric sulphide and red lead darken on
heating, and recover colour on cooling, if neither
has been heated to the point of decomposition.
However, if lime is present as an adulterant, the
discoloration remains.

Touchstone
Of all the tests in Pliny, the one which has
survived longest is the touchstone.(19) (XXXIII.2b)
"Workers skilled in its use can tell straight off
how much gold, silver or copper it contains to the
nearest scruple, by this wonderful method which
never fails them." This sounds excellent, but
Pliny has not got it quite right, because he is
referring to the testing of an ore. The touchstone
was certainly available as a test for a metal.
There is a good description of its preparation and
use fifteen centuries later in Ercker (1574).(20)

Silver
There has been a great deal of philological discussion of Pliny´s methods of acquiring his information
from written sources, but from time to time it does
seem as if he was reporting from direct observation.
Pliny must have talked to practitioners of these
metallurgical arts. He picks up frauds very
clearly as, for example (XXXIII.127) in a test for
silver. If a filing of silver is heated on an iron

shovel, it reddens if impure (that is, in modern terms, by the oxidation of any copper content), but, says Pliny, the test can be interfered with; "if the shovel is kept in human urine, the filing absorbs some while it is being heated and counterfeits brightness." (The organic matter in the urine would act as a reducing agent and reduce the oxide which would otherwise show the presence of copper.) Another test for silver, if the silver is highly polished, is to breathe on it and observe whether it fogs immediately, and "easily shakes off the dew". This has the look of good physics. The silver being of high thermal conductivity would promote rapid condensation of water vapour in breath; but so would any other conductive metal. It would not indicate the presence of any impurity in a silver alloy. It is possible that other white metals, such as lead alloys, might be passed off as silver, in which case the conductivity or dew-point test would provide a distinction between specimens tested together.

Cadmea

There is some doubt about whether zinc was ever isolated in classical antiquity. Zinc ores were known and used but in two quite different ways. Brass was known, but it was not made by direct alloying. Zinc oxide appears as the product of smelting with a zinc carbonate ore. The zinc oxide appears in various degrees of friability, depending on the part of the furnace from which it was collected. Pliny points this out (XXXIV.100). "The name **cadmea** is used to indicate more than one substance - firstly the actual mineral from which the metal is prepared, indispensable to the copper smelter and not without its use for the physician; secondly a different substance formed in the smelting furnace." Pliny distinguishes in the paragraphs which follow between **capritis, botryitis, plastitis** and **orchyitis** all of which are zinc oxide (the original ore being carbonate). All **cadmea**, says Pliny, comes from the furnaces of Cyprus, and is used for medicinal purposes. The tests for quality are all rough and ready physical tests of apparent density and friability.(21)

In medicinal use a salve was made of **cadmea** which had been reheated with charcoal, pulverised and macerated until "the consistency of the material resembles that of **cerussa** (white lead) and no grittiness can be detected by the teeth". This test, of grinding with the teeth, is also used for verdigris (a basic copper acetate). In this way, crude adulterants such as powdered pumice could easily be detected, but not one other important adulterant, namely ´shoemakers´ black´.

Atramentum

´Shoemakers´ black´ is **atramentum** or ferrous sulphate: green vitriol. The colour would be similar, but the **atramentum** would be very deleterious when the verdigris was used as a pigment or as an eye salve. To test for the presence of **atramentum**, the adulterated verdigris is heated on an iron shovel. Pliny says that if the material is pure it keeps its original colour (green) but if it contains **atramentum** it become red. As Bailey points out, this is highly suspect. Certainly the ferrous sulphate will decompose to give a red iron oxide. But so also will the basic copper acetate decompose to give first a red cupric oxide, then copper by reduction by the acetone given off from the acetate, then a black copper oxide. Pliny seems to have been quoting Dioscorides whose account of this reaction recognises the red colour of the iron oxide, but does not claim that this distinguishes between **atramentum** and verdigris. Pliny is thus interpreting as a distinguishing test something which his source described only as an individual property.

The other test for ferrous sulphate is, however, sound and historic.(22) Papyrus soaked in an infusion of gall-nuts turns black in the presence of shoemakers´ black. Kopp (in 1843) described this as "the first test-paper known" and the reaction giving the black colour was to be the basis of permanent inks (other than carbon black inks) down to modern times. Black is black in most contexts, but one is constantly in difficulty in Pliny, as in all chemical studies before modern times, over the identification of other colour references. In book XXXIV.124,125, a colour distinction is referred to which typifies the difficulty one must experience in understanding what tests if any were relied upon. Take **chalcanthus**, which is derived from the exposure of pyrites to air and water. Pliny says the **chalcanthus** made in this way is blue and may be mistaken for glass. This seems like a description of copper sulphate, which would be the principal product of the weathering of iron pyrites. Where there was copper present, or where a comparable weathering of copper minerals took place, a blue copper sulphate would be formed. The strikingly glass-like appearance of these sulphates of metals led to them receiving the general name of vitriols in later centuries. Pliny says "so a double distinction is sometimes made between the natural and the artificial substance, the latter being paler, and as inferior to the former in quality as it is in colour." It is possible that the difference observed is the result merely of the physical form, a good crystal of either copper sulphate or ferrous sulphate being darker in colour than a powder. The distinction between the two as being better or worse is not just a matter of quality. **Chalcanthus** (copper sulphate) would not be used at all as shoemakers´ black, **atramentum**, ferrous sulphate, so that we can only say that if a choice of quality was not as crude an affair as appears here, Pliny was not aware of, or did not appreciate, the significance of certain skills in craftsmen´s judgements.

Various

Pigments are the subject of a variety of tests; just how they were applied must be a matter of conjecture. No doubt individual experience applied to a restricted range of materials enabled some tests to give good results. Indigo, for example (XXXIV.46), "is tested on hot charcoal, for pure indigo gives a flame of a fine purple colour and as the smoke rises a smell of the sea may be noticed, which accounts for the popular belief that indigo is gathered on the rocks". How Pliny came to describe the colour as being "as of the sea" is a mystery, but it is true that pure indigo would volatilize easily, while adulterated indigo would not or would leave a definite residue.

Some tests, although described very circumstantially are very hard to comprehend now. There is described (XXXV.19) a test for the quality of "Eretrian earth" (which seems to have been a kind of magnesia). If Eretrian earth was rubbed on a piece of copper it produced a violet mark if of good quality. We cannot now find any substance which will do this. Bailey suggests that if a white substance is rubbed on unburnished copper, a streak might look purple.(23) He does not remark that Pliny´s **violaceus** might mean any floral colour, and that the bright colour produced by local cleansing of the surface might be sufficiently reproducible in skilled hands to constitute a trade test.

We should also reflect that colour changes were of interest to other people, and that the origins of some alchemical terminology are to be found in the "iosis" met in alchemical literature later on.

I mention only one other test which occurs in the Leyden/Stockholm papyrus as well as in Pliny, to emphasise the need for skill. Tin is tested by pouring molten tin on to papyrus.(24) It should seem to burst the papyrus by its weight rather than by its heat (XXXIV.163). However, as Caley and Halleux point out in their discussions of the Leyden papyrus, a eutectic of lead and tin would have a lower melting point than either and so pass this test.

The presence of tin in the lead would not be detected. But what was the tester looking for? Clearly, utility. The ultimate test was effectiveness in use, and we should therefore look at all these testing procedures in terms of general practice.

What emerges from this selection of tests, often misunderstood by Pliny, and often obscure, even with our chemical knowledge, is a widespread habit of systematic discrimination, subject to a consensus of method and interpretation. This is by no means to be taken for granted in any period or community. It is easy to observe excellence of workmanship in surviving artifacts or to note the endurance of monuments, and to infer the skill of individual craftsmen or designers. What is less obvious is the way in which continuity of practice was ensured. In the tests I have mentioned we have seen several factors bearing on this continuity.

1. There was only a very limited technique of identification of a substance of unknown composition. If the history of a specimen was unknown, there was no means of finding out, for example, whether a specimen of a greenish crystalline substance was **atramentum** or impure sea-salt. Even taste was unreliable in such a case. To determine quality one had to have a good idea of what the substance purported to be. However, one has to remember that very few analytical problems nowadays present the analyst with a substance about which he knows nothing. The forensic analyst is sometimes in this position, but in industry or research the chemist is generally looking for variations from a norm. The craftsman or tradesman in Pliny's day was generally buying what he had before, from a known source, and would therefore be concerned with satisfying himself about the maintenance of quality or protection against fraud. We can therefore say that the lack of precise means of identification

need not imply a lack of an orderly and consistent approach to testing.

2. We can note the widespread use of some tests which were of particular significance, outstanding being the tests for gold. This was not just a test for usefulness of a material for some practical application, although it did mean this sometimes, as for example in gilding. Gold and silver were essential to the conduct of commerce and the control of government, so the testing of the value of a specimen of bullion or coin was already ancient in Pliny's day and had been widely practised.

3. We see in many places the differentiation of qualities of a given product, as for example, the different varieties of haematite distinguishable with the touchstone.

Our general conclusion must be that the rule of thumb testing of materials was widespread in antiquity, and that only the humble level of its pursuit prevented our being able to elicit detailed information about skilled practices.

Pliny's literary approach probably obscured information which reached him in a much clearer form than that in which he tansmitted it. For example, the language of the Leyden papyrus is far more intelligible in some cases. Recipe 43 runs: "If you wish to test the purity of gold, remelt it and heat it; if it is pure it will keep its colour after heating and remain like a piece of money. If it becomes white, it contains silver; or it becomes rougher and harder, some copper and tin; if it blackens and softens, lead."

Taking this kind of outlook together with that of Pliny himself, we are left with the impression of life in his world which was not by any means scientific but did recognise that the material substances which we must use are susceptible of consistent and rational examination for fitness for purpose and therefore for for the making of everyday technical and economic decisions. This is rough intellectual ground but in it the seeds of science were one day to germinate.

NOTES

1. K C Bailey, **The Elder Pliny's Chapters on Chemical Subjects,** (edited with translation and notes), part 1, London 1929, part 2, London 1932

2. "Chemistry" as a clearly defined area of study is quite a recent idea, dating from the 18th century fashion for the ordering of knowledge. The

idea of a **chemist**, the practitioner in certain arts, is older. The disentangling of practical chemical pursuits from the wishful thinking of the alchemist is a constant preoccupation of historians. See for example J R Partington, **History of Chemistry,** vol.1, part 1, London, 1970, pp. xi-xviii (an introductory essay on "Chemistry as rational alchemy".

3. The literature of the history of alchemy is vast and curious, from careful, disciplined studies to wild misguided speculation. The reader approaching the subject for the first time is advised to start at the disciplined scholarly end with the volumes of **Ambix,** the journal cf the Society for the History of Alchemy and Chemistry. See also A Pritchard, **Alchemy: a Bibliography of English Language Writings,** London, 1980

4. The reader who has not yet met the early alchemical literature in its more florid forms could do no better than read the Book of Revelation, the tone and manner of which have the flavour of Hellenistic alchemy, if not demonstrably its intention. Sherwood Taylor gives enough lengthy quotations from several writers to show the alchemical attitude to nature and to imagery at several periods. F Sherwood Taylor, **The Alchemists,** London, 1952

5. R P Multhauf, **The Origins of Chemistry,** London, 1966, summarises the position before and around Pliny´s time.

6. One example will be enough to illustrate the vast scope of the subject. Kolthoff and Elving´s **Treatise on Analytical Chemistry** (a cumulative work of the third quarter of the 20th century) has reached 26 volumes for its parts 1 and 2. Part 3 is in progress and likely to be equally massive.

7. F Greenaway, **Studies in the Early History of Analytical Chemistry,** MSc Diss., University of London, 1957

8. E R Caley, **J. Chemical Education,** 3, 1926, pp. 1149-1166; and 4, 1927, pp. 979-1992 is the English language translation most commonly quoted. The first introduction of the subject to modern historiography was the translation in M Berthelot, **Introduction a l´étude de la chèmie des anciens et du moyen age,** Paris, 1889

9. R Halleaux, **Les alchemistes Grecs (1)** Payrus de Leyde, **papyrus de Stockholm, fragments et recettes. Text établi et traduit par Robert Halleux,** Paris, 1981. This is the most recent study of the Leyden/Stockholm papyrus with complete Greek text.

The translation differs from Caley in some respects, but there is nothing which strikingly changes our view of the nature and origin of the material.

10. Lynn White, **Medieval Technology and Social Change**, Oxford, 1962. The several sections illustrate the different speeds at which different kinds of technical innovation produced social effects.· White´s period is later than ours here but his method of approach is relevant.

11. M P Crosland, **Historical Studies in the Language of Chemistry**, London, 1962

12. F Sherwood Taylor, **J. Hellenic Studies** 50. 1930, pp. 109-139

13. A R Williams, "The production of saltpetre in the middle ages", **Ambix** 22, 1975, pp. 125-133

14. F Greenaway, "Historical continuity of the tradition of assaying", **Ithaca 26 viii - 2 ix 1962** (Proceedings of Tenth International Congress of Science) pp. 819-823

15. J H Craig, **Newton at the Mint**, Cambridge, 1946

16. O Vittori, "Pliny the Elder on gilding", **Gold Bulletin**, 12, International Gold Corporation, Marshalltown, South Africa.

17. See Crosland, n.11 above, passim.

18. M E Weeks, **Discovery of the Elements** (rev. H M Leicester), Easton, Pa (**J. of Chemical Education**) gives numerous references to antimony (mainly occurring as sulphide) in ancient times: pp. 95-103

19. See Greenaway, n. 7 above.

20. L Ercker, **Beschreibung Allerfurnermisten Mineralischen Ertzt und Berckwercks Arter...**, Prague, 1547. (Most conveniently consulted as **Lazarus Ercker´s Treatise on Ores and Assaying**, translated from the German edition of 1890 by A G Sisco and C S Smith, Chicago, 1951, with introduction and notes.

21. See Weeks, n. 18 above, under references to zinc, passim.

22. M Nierenstein, **Isis** 6, 1931, pp. 43-46

23. F S Taylor, **J.Hellenic Studies.**

24. Caley and Halleux, nn. 8 and 9 above.

Chapter Ten

SOME ASTRONOMICAL TOPICS IN PLINY

O Pedersen

Introduction

Does the fact than an ancient Latin author wrote
extensively on astronomical matters entitle him to a
place in the history of astronomy? This is a
central question with regard to Pliny the Elder,
whose immense **Naturalis Historia**(1) has been justly
praised by successive generations of scholars as an
"inexhaustible source of information on the scientific knowledge and social conditions of the early
Roman imperial period". On the other hand, as far
as astronomy is concerned it has become more and
more clear that "Pliny had no real understanding of
this topic" because his "uncompromising collector's
attitude prevented him from studying technical
problems seriously" at the same time as his "honesty
and human warmth shines through his piles of
slips".(2) Considering the scope and extent of the
Natural History, these piles must have formed a very
impressive collection. Pliny himself reckoned that
he had compiled twenty thousand memorable facts from
two thousand books by one hundred different authors
(Preface, 17)(3) even if he restricted himself to
what he called **sterilis materia, rerum natura, hoc
est vita** (Preface, 12-13), finding no place for
digressions, speeches, discourses, marvellous cases
or mere **faits divers,** although such matters might
have been interesting to tell or pleasing to read.
No one can blame him that he was unable to stick to
this strict discipline.

Thus Pliny's work was conceived on a larger scale
than other works belonging to the encyclopedic genre
which the Romans had learned to imitate from the
Greeks with the (now lost) **Libri IX Disciplinae** by
Marcus Terentius Varro (116 - 27 BC) as the most
outstanding example. It was also marked by a
different attitude. Pliny wished to achieve

162

something which nobody before him had ventured upon,
neither among the Romans nor among the Greeks
(Preface, 14). In the traditional Greek **paideia**
astronomy had found its place together with
arithmetic, geometry and theory of music within that
section of the liberal arts which became known later
as the **quadrivium.** The structure and contents of
the **Natural History** reveal that Pliny consciously
rejected the liberal arts as an adequate framework
of human knowledge. These subjects had become
disciplined, specialised and carried a high degree
of abstraction. Accessible to scholars, they were
difficult to grasp for ordinary people of the world
with practical rather than theoretical needs.
Moreover, they contained only a small part of all
available knowledge, ignoring most of the manifold
phenomena of the natural world. Accordingly, Pliny
decided to concentrate on subjects of more immediate
importance for human life in general. One
consequence of this deliberate choice was that
disciplines like arithmetic, geometry and harmonics
were discarded in favour of subjects more directly
related to everyday life such as the geography of
places where man lives snd works, and the zoology
and botany of the animals and plants which surround
him and provide him with the most urgent necessities
of his existence - shelter, food, drink and
medicines. All these fields of knowledge had their
own literature. But Pliny was right - nobody
before him had ever ventured to survey and expose
them all within the scope of a single work.
    Against this background it becomes a question why
astronomy survived and was given such a prominent
place in the **Natural History** at the expense of the
other liberal disciplines. If we ignore the
bibliographical book I, the **Natural History** opens
with an extensive, general account of the universe
comprising about one half of book II, just as there
are long sections on astronomical matters, or other
subjects presupposing astronomical knowledge, in
books VI, VII, VIII and XVIII. This particular
status of astronomy has to be explained.

The rejection of Astrology
In Hellenistic times the theoretical description of
celestial phenomena had become more and more
involved with astrological forms of divination. By
the time of Pliny this wave had reached the Roman
world, supplanting or at least supplementing those
earlier methods of divination which Cicero had

described in the **De Divinatione** and which, mixed with magical procedures, were known as the **Etrusca disciplina,** although the Romans were perfectly well aware of the ´Chaldean´ origin of astrology as such.

Orginally closely connected with Mesoptamian astral religion and its belief in the planets as deities governing the world, astrology had to some extent been secularised by the Greeks and transformed into the semblance of a rational science which in the century after Pliny found its most perfect expression in the **Tetrabiblos** of Ptolemy. The stars and planets still ´governed´ things here below, but they themselves were governed by inexorable laws of nature which astronomers would try to disclose. Once these laws were known it would be possible to predict future positions of the planets and, in consequence, their future ´influences´ on terrestrial and human phenonema.

This new attitude appealed to people who were at the same time convinced of the possiblity of divination as such and of the new idea that the laws of planetary behaviour must be shaped in a mathematical form. Thus astrology and ´mathematics´ became more and more synonymous, a tendency which is apparent already in Cicero and later gave rise to St Augustine´s fulminations against the impious ´mathematicians´. It would seem to be a likely hypothesis that as a child of his time Pliny retained astronomy within his account of nature as a necessary prerequisite of the ´science´ of astrology. But this explanation is clearly untenable.

Here it is important to remember that Pliny was not opposed to divination as such. The **Natural History** is full of reports of omens and portents culled from numerous historical annals and travellers´ tales. Some of these accounts are dismissed as superstitious, while others are more or less accepted. However, in the special case of astrology, Pliny´s sceptical attitude prevails. Thus he definitely rejects the old astral religion which was the cradle of astrology. Jupiter and Mercury are not gods and in no way comparable to a supreme Being. Moreover, if such a supreme Being exists it is a ridiculous notion that He cares about human affairs (II.5.20). It is also impossible to place a deified **Fortuna** between God and men, since this would place us at the mercy of chance (II.5.22). Furthermore, Fortune cannot be replaced by attributing events to the influence of the stars; this would mean that God´s decree had been enacted

once and for all since the stars themselves are
governed by laws. (II.5.23). Pliny is aware that
astrology is everywhere on the increase among the
learned and unlearned. Nevertheless, it remains a
superstition since there is no such close connection
between us and the heavens that our mortality is
shaped by the stars as Pliny says with reference to
the belief that a falling star signifies the death
of a human being (II.6.29).

This latter quotation points to Pliny's most
concerted attack on astrology as we find it in book
VII which deals with all the aspects of the **conditio
humana**, including the possible lifespan of individual human beings. Here Pliny notices that "this
topic seems of itself to call for the views held by
the science of the stars" (VII.49.160) since this is
the opinion of many philosophers, among whom Pliny
mentions Berosus the Chaldean, the Byzantine astr-
ologer Epigenes (who was also known to Posidonius
and Seneca), Aesculapius the father of medicine, and
the "ancient Egyptians" Nechepsos and Petosiris.
Against this well established although non-Roman
tradition, Pliny takes refuge in two incontro-
vertible facts. First he demonstrates that astr-
ologers often disagree in their prognostications so
that "the variation in the science shows how
uncertain it is" (VII.49.162). Secondly, he app-
eals to the empirical evidence provided by the stat-
istics of a census of population carried out under
his own emperor, Vespasian. Restricting himself to
data from the middle region between the Apennines
and the river Po he is able to show the great
variability in the length of human life: people die
at different ages even if they are born 'under the
same star' (VII.49.165). Finally, historical
records point in the same direction. Thus we read
in Homer that Hector and Polydamus were born in the
same night although their individual fates were very
different. This was the case also of the two Roman
orators Marcus Caelius Rufus and Gaius Licinius
Calvus. In short, taking the entire world,
this happens daily even to persons born at the same
hour - masters and slaves, kings and paupers come
into existence simultaneously (VII.49.165) - an
argument within the spirit of the **De Divinatione** of
Cicero(4) and later utilised with great effect by St
Augustine.(5) After this it is not surprising that
Pliny carries a particularly violent attack against
the progress of astrological medicine in his own day
as practised, for instance, by Crinas of Marseilles,
who "united medicine with another art and regulated

the diet of patients by the motions of of the stars
according to the almanac of the astronomers",
leaving an immense fortune behind him when he died.
He behaved almost as badly as his townsman Charmis
whose patients were used as guinea-pigs for his
experiments with cold baths. "There is no doubt"
says Pliny, "that all these, in their hunt for
popularity by means of some novelty, did not
hesitate to buy it with their own lives"(XXIX.5.9).

God, Man and the Universe
Whereas Pliny rejected astrology he was much more
favourably disposed towards another line of thought
which found some kind of justification of astronomy
in the spiritual values inherent in the study of the
universe. That the things on high draw the mind
and soul of man upwards from this base world was a
theme dear to many ancient philosophers at least
from the time of the Pythagoreans. It was
propagated by both Plato and Aristotle and developed
with Stoic overtones in the preface to Ptolemy´s
Almagest. Later it became one of the basic ideas
of the **De Consolatione Philosophiae** of Boethius,
from which it was adopted by the middle ages where
it found its most striking expression in the speech
of Theseus towards the end of **The Knight´s Tale.**
In this tradition Pliny has an important place.
Right at the beginning of book II he abandons his
programme of providing only factual information in
order to preach a kind of sermon on the subject of
´God and the universe´ in which the world as a whole
is depicted as sacred, immense, finite yet similar
to the infinite, and eternal and self contained so
that it would be madness to ask what lies outside
the univérse as long as our knowledge of the things
inside it is so imperfect. Also it would be vain
to speculate on the possible existence of other
worlds than ours (I.1.1-5).
This majestic opening passage is clearly meant to
instill a wholesome awe of nature in the mind of the
reader and Pliny has not written many lines before
he confesses that to him the world is fitly
conceived as a deity - **numen esse credi par est**
(II.1.1) - but a deity which ·surpasses human
understanding. Nevertheless, we understand enough
to begin to realise that we must reject a number of
ideas about God which are current in popular
religion. Thus it would be foolish to ask for the
form and shape of God, for "whoever God is, if He is
different from the world",(6) He must be wholly

sense, sight and hearing, wholly of soul, wholly of
mind, wholly of Himself (II.5.14). It would be an
even greater folly to speak of gods without number
and also of deified human virtues as when people
speak of particular goddesses of mercy, hope,
concord, intelligence **et cetera**. Popular poly-
theism, with all its human gods marked by moral
weakness and shortcomings, comes near to childish
fantasies - **puerilium prope deliramentorum est**
(II.5.17). For whatever Pliny´s God is, He is
morally perfect and, in the last resort, the same as
the moral order of the world. Accordingly,
divinity equals true morality, and God is that
mortal aids mortal - **Deus est mortali iuvare
mortalem**(II.5.18). In consequence it is only
proper that men should deify the highest exponents
of morality such as the Roman emperors and in
particular "the greatest ruler of all times, His
Majesty Vespasian, who came to the rescue of an
exhausted world" (II.5.18) - an interesting example
of how a theological conception of the universe
might lead a stoic mind to acceptance of the new
imperial cult.
    If Pliny´s God is a transcendant, morally perfect
Being, He is not omnipotent but subject to several
limitations. Thus He cannot commit suicide,
although He had bestowed this supreme privilege upon
man. Neither can He make man eternal, nor recall
the dead. Furthermore He is unable to undo what
has actually happened, for instance by causing a man
who has lived not to have been alive. Finally He
cannot cause ten plus ten to make anything else
other than twenty. All this, says Pliny, just
"shows the power of nature and its identity with
what we call God"(II.5.27), adding that this
digression has not been irrelevant since the
unending controversy about God has made the subject
familiar to everyone. Thus it is possible to speak
of a theological justification of astronomy in
Pliny, since it is obvious that the study of the
universe cannot but purify human ideas about God.
    This has a corollary in the purely human realm.
If God is One there are no separate deities behind
the phenomena of nature and we have no reason to
regard them with irrational feelings of fear.
Pliny makes much of the story of how a lunar
eclipse was predicted by the Roman general Sulpicius
Gallus just before the battle of Pydna in 168 BC
with the result that the Roman army was liberated
from fear and was able to win a victory. Pliny
believed that Hipparchus had found general methods

for such predictions and takes this as his text for another sermon in praise of the almost super-human heroes of science:

> men of more than mortal state, who have dis-
> covered the laws of so many deities (that is,
> celestial bodies) and released from fear the
> miserable mind of man who now regarded with
> horror eclipses as some sort of crime or death
> of the stars... now regarded the dying moon as
> the victim of poisoners and came to her rescue
> by making discordant noises... Praised be your
> genius, you who interpret the heavens, comprise
> all nature and discover laws which vanquish
> gods and men! (II.9. 53-54).

Eclipses are perfectly natural phenonena and can be predicted. One has nothing to fear from them. The whole scientific revolution brought about by the Greeks is summarised in this insight, but rarely expressed so clearly as in this passage by Pliny.

The Practical Purpose of Astronomy
The destruction of the primordial fear of nature and the knowledge of God as the natural order of the universe are not the only reasons for studying astronomy. They are not even the most important. For as the heavens belong to God, so the earth belongs to man - **Sic hominum illa ut coelum dei** (II.53.154). And Pliny is above anything else a man of the earth who devotes almost half of the second book of the **Natural History** to the earth sciences and all the rest of his work to a description of things here below. It is not for nothing that the sermon on ´God and the Universe´ is paralleled by a long, poetical and highly emotional discourse in which the earth is praised as the only part of the world which is venerated by the name ´mother´ since "she receives us at birth, and gives us nurture after birth, and when once brought forth she upholds us always, and at the last when we have been disinherited by the rest of nature she embraces us in her bosom and at that very time gives us her maternal shelter".(II.53.154)

In more prosaic language this means that we live by the fruits of the earth and that agriculture must be the most fundamental human occupation, as decribed in book XVIII of the **Natural History**. But this industry, says Pliny, "is to a large extent connected with astronomy, and we will begin by

setting out the views of all authors in regard to
it" (XVIII.55.201). Accordingly he quotes Zoroaster for his advice to farmers to sow when the sun
has passed the 12th degree of Scorpius and the moon
is in Taurus. On the other hand, in his **Praxidike**
Attius maintains that the moon must be in Aries or
in Gemini, Leo, Libra or Aquarius (XVIII.55.200),
while Hesiod used the Pleiades as indicators of the
correct time of sowing (XVIII.57.213). Since there
is disagreement among the authorities, the matter
must be difficult and great care is necessary.
Virgil is quoted for saying that correct farming
must be based on observations of the stars and the
winds which are just as precise as those made at sea
for the purpose of navigation. Since farmers are,
in general, unfamiliar with astronomy Pliny must
consider it "an arduous and vast aspiration, to
succeed in making the divine science of the heavens
known to the ignorance of the rustic - but it must
be attempted, owing to the vast benefit it confers
on life"(XVIII.55.206).

This was, perhaps, the most important reason why
Pliny included astronomy in his encyclopedia.
Civilised life presupposes agriculture, and agriculture presupposes astronomy since the successful
farmer would have to observe a whole range of
celestial phenomena as part of his daily routine.
In consequence it would be a mistake to examine
Pliny´s astronomy only on the basis of the
theoretical chapters of book II, since he himself
seems to have attached more real importance to the
practical application of astronomy as displayed in
chapters 55 to 75 of book XVIII. This latter
section falls in two parts, in the first of which
Pliny reflects on the difficulties of applying
astronomy to practical purposes (chapters 55 to 59)
while in the second part (chapters 60 to 75) he
submits a ready made astronomical calendar of
immediate use to the farmer.

The problem is if it is possible to connect the
annual rhythm of agricultural work with astronomical
phenomena which can be easily observed and thus
serve as indicators of the correct times for sowing
and reaping and what else the farmer has to do. To
solve this problem one has to proceed in an
empirical way comprising two different stages.
First one must discover the laws of the relevant
phenomena, and next apply them to specific evidence;
**Primum omnium a caelo peti legem, deinde eam
argumentis esse quaerendam** (XVIII.57.210).

However, this programme is not carried out until

Pliny has discussed the possibility of a simpler, non-astronomical method. If it were possible to regulate the annual work of the farmer simply by the four seasons as they are indicated in an ordinary calendar one might do away with astronomy altogether. But on this point he reveals some rather strange misgivings stemming from the fact that the calendar introduced by Julius Caesar assisted by Sosigenes (XVIII.57.211-212) operates with a solar year of 365 and a quarter days, obtained by letting three common years of 365 days be followed by a leap year of 366 days. This means that it is impossible to define precise times of the stars - **certa siderum tempora** - by the calendar since the calendaric seasons may begin sometimes before and sometimes after the precise times given by the positions of the fixed stars (XVIII.57.207). This is not very clear, but may refer to the fact that the dates of the equinoxes and solstices may vary one day within the four year period. This is true, although it is strange to think that one day more or less would seriously affect practical agriculture. Nevertheless, it seems important to Pliny that the vernal equinox always is on the same date, that is, **ante diem viii kal. Aprilis,** or March 25 (XVIII.66.246). This is the traditional Roman date of the beginning of the spring which was later to cause so much trouble between the Romans and the Alexandrians with respect to the ecclesiastical calendar and the Easter Compotus(7). Similarly Pliny is worried by the fact that the lengths of the astronomical seasons cannot be expressed by integral numbers of days, quoting Hipparchus for the values of 94 days 12 hours and 88 days 3 hours for the longest and shortest season respectively. Finally he remarks that the four year period of the seasons in the Julian calendar is not the only one. There is also an eight year period corresponding to 100 revolutions of the moon and eight revolutions of the sun (XVIII.57.217). This is a badly digested reference to the old **octaëteris** period stating that there are 99 synodic months in eight years That this luni-solar cycle has nothing to do with the seasons reveals that Pliny is no great authority on calendaric matters.

The Farmer´s Calendar
Since according to Pliny the farmer would be badly served by the ordinary calendar it is impossible for him to do without paying close attention to such astronomical phenomena as are suitable for his

purpose.   Here Pliny follows the age-old tradition
of the Greek **parapegmata,** or farmers´ calendars,   in
which the course of the solar year is related to the
annual   disappearance   and reappearance of the fixed
stars(8).   The sun performs an annual motion around
the heavens from west to east.   When it overtakes a
fixed star the latter is invisible since it is above
the horizon together with the sun and rises and sets
at the same times.   But as the sun moves further to
the east the star is, as it were, left behind,   and
on   a   certain   date   its   distance from the sun has
increased so much that it will be visible   over   the
eastern   horizon   just   before   sunrise.   This
reappeareance of   the   star   after   its   period   of
invisibility is usually called its ´heliacal rising´.
In something less than   a   year   this   situation   is
reversed.   Now the sun approaches the star from the
west   so   that   it becomes once more invisible.   On
the last day of the visibility period, the   star   is
seen   for   the   last   time above the western horizon
just after sunset.   This is its ´heliacal setting´.
Pliny describes these stellar phases in   some   detail
(XVIII.58.218-9)   with   the sensible remark that the
terminology is   misleading   and   that   it   would   be
better   to replace ´rising´, **exortus** by ´emergence´,
emersus, and ´setting´, **occasus,** with ´concealment´,
occultatio.   This is only a single instance of   the
difficulties   met   with   by   the   Romans in rendering
Greek technical terms into Latin.(9)
     Pliny´s **parapegma** is like other works of the same
kind,(10) based on   a   great   number   of **parapegmata**
culled   from   a   variety   of   authorities among whom
Virgil and Varro(11) play a conspicuous role.   But
he   also   quotes **parapegmata** of   both   Egyptian and
Babylonian origin   of   which   little   or   nothing   is
known   from   other   sources.   Thus   he is the only
authority   for   the   existence   of **parapegmata** from
Babylon (XVIII.57.211)(12).   The actual compilation
of   the complete list gave rise to a new worry about
the   very   possibility   of   using   astronomy   in
agriculture, since it clearly revealed disagreements
of   the   authorities.   Thus   Pliny   mentions   that
Hesiod placed   the   heliacal rising of the Pleiades
(indicating the correct time of reaping) just   after
the   autumnal equinox, whereas it was placed 25 days
after   the   equinox   by   Thales,   30   days   after   by
Anaximander,   44   days after by Euctemon and 48 days
after by Eudoxus   (XVIII.57.213-4).   Nevertheless,
that   did   not   prevent   Pliny   from   making   use of
traditional information in   the **parapegma** which he
published   in   chapters   60   to   75   of   the **Natural**

Astronomical Topics

History and which we must refrain from describing in
detail.   A single quotation will show the general
character of the work.   Thus we read about May and
June that

> After the rise of the Pleiads the weather is
> indicated by Caesar by the morning (that is,
> heliacal) setting of Arcturus on the following
> day, the rise of the Lyre on May 13, the
> setting of Capella, and in Attica of the Dog,
> in the evening of May 21.   On May 22 according
> to Caesar, Orion´s sword begins to set, and in
> the evening of June 2 according to Caesar and
> also for Assyria the Eagle rises.   On the
> morning of June 7 Arcturus sets in Italy and on
> the evening of June 10 the Dolphin rises.   On
> June 15 Orion´s Sword rises, but in Egypt this
> takes place four days later... while on June 24
> the longest day and the shortest night of the
> whole year make the summer solstice.   In this
> interval of time the vines are pruned, and care
> is taken to dig around an old vine once and
> around a new one twice.   Sheep are sheared,
> lupins are ploughed in to manure the land, the
> ground is dug over, vetches are cut for fodder,
> beans are gathered and then threshed
> (XVIII.67.255-7).

Since the work of the farmer depends on the weather
as well as on the time of year, it is not surprising
that Pliny includes a long section on astronomical
meteorology in the **Natural History** (XVIII. 68-70)
supplementing what he had to say on meteorological
matters in book II (II.38.102-61, 153).   His
purpose is to prove how the weather is influenced by
the sun, moon and stars, and in particular how these
heavenly bodies influence the twelve classical
winds.   We notice here one of the few cases in
which he gives a detailed description of a scientific instrument in the form of a windrose which may
be drawn on the ground on simple astronomical
principles (XVIII.66.326ff.) or carved on a block of
wood (XVIII.67.331-9) as a useful tool for the
farmer for ascertaining the direction of a
particular wind which can thus be identified.
When Pliny compiled his calendar he had no single
geographical locality in mind.   He explicitly drew
attention to the fact that the heliacal rising of
the constellations must happen on the same date in
all places with the same geographical latitude
(XVIII.57.217).   He does not explain how the calen-

dar can be adapted to different latitudes, but
nevertheless concludes the geographical section of
the **Natural History** with a detailed account of a
"theory discovered by the Greeks and showing much
subtle ingenuity" (VI.39.211), that is, how the
earth can be divided into a number of segments which
"we call circles and the Greeks parallels"
(VI.39.212). These segments are the famous seven
classical **climata,** or ´climates´ each of which is
characterised in three different ways. The first
climate is defined in the first instance by the
regions it contains, that is, Persia, Mesopotamia,
Alexandria, Carthage and the Pillars of Hercules, to
mention but a few of them.

Secondly, everyhere in these regions the
equinoctial noon shadow of a vertical gnomon seven
feet high will be four feet long, and thirdly, the
longest and shortest days here will be 14 and 10
equinoctial hours respectively. The other six
climates are dealt with in the same way.(13) The
whole system can be summarised in the following
table, where g is the height of the gnomon, So the
length of the equinoctial shadow, and M the longest
day in equinoctial hours.

Climate	g	So	M
I	7 ft	4 ft	14
II	35 ft	24ft	14 2/5
III	100inches	77inches	14 8/15
IV	21 ft	16ft	14 2/3
V	7 ft	6 ft	15
VI	9 ft	8 ft	15 1/9 or 1/5
VII	35 ft	36ft	15 3/5

Pliny vaguely refers to "the ancients" who have
worked out this scheme, and to "subsequent students"
who have added to it by introducing a parallel with
a 16 hour day passing through Borysthenes (the mouth
of the Dnieper), another with M=17 hours through
Britain, and finally a parallel through Thule "where
there are alternate periods of perpetual daylight
and perpetual night" (VI.39.219). These anonymous
authorities have also added two more parallels to
the south corresponding to to M=12 1/2 and 13 hours.

These Plinian **climat**a have given rise to much
discussion and there is no doubt that the text is
more or less corrupt.(14) We shall not here go
further into this problem, but only notice that
Pliny does not give geographical latitudes of the
climates in degrees, and that he obviously had no

understanding of the mathematical relationship between latitude, gnomon shadow and longest day which
Hipparchus had mastered a couple of centuries
earlier. Mathematical geography and its spherical-
astronomical presuppositions were clearly beyond his
grasp.

The Structure of the Universe

Pliny´s conception of astronomy as more of a
practical necessity than a liberal art entails two
important consequences. Firstly, it explains why
his introductory account of astronomy and cosmology
in book II has rather a selective character.
Secondly, we would be wrong in regarding this
account as a testimony to the general state of
astronomy around the middle of the first century.
The astronomy of Pliny appears as a highly personal
affair outside the main stream of classical
astronomical thought and - with a single exception -
without any pretension of new or original
contributions.

The principal features of Pliny´s universe are
in general agreement with Hellenistic cosmology.
The world or **mundus** is supposed to be spherical and
thus perfectly adapted to its daily rotation (II.2.5
and 3.6) which produces the rising and the setting
of the sun and the stars. It is not easy to say
whether it also produces any ´music of the spheres´
unheard by human ears; nevertheless, Pliny
faithfully records the Pythagorean application of
the theory of music to the problem of the mutual
distances of the planets, calling it a "subtility
more entertaining than convincing"(II.20.84).

The sphere of the world contains several separate
regions. At its outer boundary are the fixed
stars. They are arranged in constellations, the
number of which is not clearly stated. In one place
Pliny speaks of "countless figures of animals and
things of many kinds" (II.3.7), but elsewhere he
mentions exactly 72 "signs" defined by the experts
and containing a total of 1600 stars (II.51.110).
The origin of this number is unknown. Later in his
Almagest (VII.5-VIII.1) Ptolemy published a catalogue of 1026 fixed stars arranged in 21 northern,
12 zodiacal and 15 southern constellations. It is
possible that Pliny here confused the 12 zodiacal
constellations with the 36 Egyptian ´decans´, adding
to the latter the 36 extra-zodiacal constellations
known from Ptolemy.(15)

Pliny´s ideas on the nature of constellations are

174

somewhat puzzling. Like many other authors he wrongly derives the word **coelum** (heaven) from **caelum,** which means the tool of an engraver, with the consequence that the constellations are real images engraved on the surface of the world, which is, therefore, not smooth but uneven, in contradiction to what "very famous authors have said" (II.3.7). But in the same place he also asserts that when the "seeds" of the stars fall into the sea they "generate monstrous effigies" - as if the constellations were living things able to reproduce themselves by some kind of semen. Here Pliny seems to rely on the ancient mythic or even animistic conception of nature which Greek astronomy had abandoned long before his time.

The remaining part of the universe is said to be the realm of the four elements. Topmost is the element of fire which is said to be responsible for the light of the stars. Next follows the air which is the "spirit of life" and penetrates all the things in the whole universe; clearly Pliny's "air" has much in common with the Stoic **pneuma.** It is also the force of the air which keeps the water and the earth suspended in the middle of the world (II.4.19). Thus a stable system is formed since "the light bodies are prevented by the heavy ones from flying up, while on the contrary the heavy bodies are prevented from tumbling down by the upwards tendency of the light ones" (II.4.11), so that each body remains stationary in its own place. However, in the same passage he also ascribes the stability of the elementary world to the unceasing revolutions of the universe around it. Finally the earth is said to remain where it is because nature denies it any other place since the rest of the world is filled with other elements (II.65.162). All this is rather vague, and it seems difficult to believe that Pliny had any real understanding of the principles by which Aristotle had connected the physics of the elements with the structure of the universe.

According to Pliny, all men agree that the earth is a spherical globe. He explains its shape by saying that the mass of the earth swells out from the centre and is, as it were, turned into a sphere by the revolution of the world around it (II.64.160). The objection that the existence of mountains prevents the earth from being perfectly spherical is met by the assertion that that all radii from the centre to the tops of the highest mountains would define a perfect sphere (II.64.160).

After all this nonsense it is a pleasant surprise to see that Pliny proves the sphericity of the oceans by a number of reasons drawn from experience; among them is the gradual appearance of a ship approaching the coast (II.65.164).

In this cosmogological scheme we must find place for the seven planets. At first Pliny says that they move in the air and are suspended by it (II. 4.12). This is a pre-Aristotelian idea which The Philosopher had discarded in favour of his theory of the ´ether´ as the substance of the celestial world, located above the sphere of the moon. But later Pliny refers to a "frontier where the air ends and the ether begins" (II.7.48), as if the ether and fire were the same elements. Furthermore, he distinguishes between the atmosphere as a mixture of terrestrial vapours and the superior element of air, and the airy region itself, the atmosphere being situated far below the moon (II.38.102), as if the air extends itself to the moon with the consequence that the moon belongs to the "ethereal" region. All these statements make it difficult to say whether or not Pliny knew and acknowledged the Aristotelian distinction between a sublunary, elementary world and a supralunary ethereal region.

One further general feature of Pliny´s universe is that it is eternal (II.1.1). On this point he is in agreement with Aristotle and there is nothing left of the oriental idea introduced by the Pythagoreans and later given prominence by Posidonius of the universe as subject to an unending series of recurrent destructions and re-establishments. That he knew this idea appears from a passing reference to the **ratio anni magni** (II.6.40), the "Great Year" being the life-span of the universe between two consecutive destructions, determined by recurrent conjunctions of all the planets. Since Pliny never returns to this notion we may conclude that he adhered to what might be called a ´linear´ conception of cosmological time.

The System of the Planets
As one might expect, Pliny knows the seven ´classical´ planets which move around the earth in two ways. First, they take part in the diurnal motion of the heavens from east to west, and secondly they perform individual motions towards the east among the fixed stars. In the middle of the planets is the sun, which is described in Homeric

terms as the soul or mind of the universe
(II.4.12f.). Highest among all is Saturn (II.6.32),
then follows Jupiter (II.6.34) and Mars (also called
Hercules). Below the sun is Venus, which is called
Lucifer as a morning, and Vesper as an evening star.
Its identity is said to have been proved by
Pythagoras of Samos about the 42nd Olympiad (in 142
A U C; II.6.36). Next comes Mercury, or Apollo
(II.6.39) and the series ends with the moon, which
is nearest to the axis - **proxima cardini** - of the
world (II.6.41-48). Thus Pliny adheres to that
order of the planets which had become canonical long
before his time, although without any real proof.
It rests on the assumption that for example Saturn
must be further away than Jupiter since it moves
more slowly (II.6.34). The periods of revolution
of the individual planets are given with the values
shown in the table.

Periods of revolution

Saturn	30	years
Jupiter	12	years
Mars	c.2	years
Sun	365 1/4	days
Venus	348	days
Mercury	339	days
Moon	27 1/3	days

The problem is what kind of periods they are.
Pliny says that Saturn uses 30 years to return to
its **sedis suae principiae** (II.6.32) and the same
expression is used of the sun (II.6,45) while Venus
peragit signiferi ambitum in 348 days (II.6.38).
Here **signifer** means the zodiac so that we can
conclude that Pliny wished to indicate zodiacal
periods of the ´motion in longitude´ - an expression
he never uses. But it is impossible to see whether
he has sidereal periods (of return to the same fixed
star) or tropical periods (return to the vernal
equinox) in mind. Presumably he did not
distinguish between them since there is no evidence
that he was aware that already Hipparchus had
discovered the phenomenon of precession, which is
responsible for the slight difference between the
two kinds of zodiacal periods.
    That Pliny had no real understanding of the
motion in longitude appears from what he says about
the sun, the course or **meatus** of which is divided
into 360 **partes** with the consequence that one has to

177

add 5 1/4 days to the year in order that the shadow cast by the sun can revert to the starting point (II.6.35). Even if the text may be corrupt it seems that Pliny assumes a daily motion of the sun of one degree and thus gets into trouble with the length of the year.

Pliny is aware that the planets also perform what we now call a ´motion in latitude´, but is unable precisely to describe it. All he can say is that the planets sweep across a greater or lesser part of the **signifer**, which is conceived as a zone or belt containing the zodiacal signs, with a width of $2°$ (II.13.66). He mentions that Anaximander was the first to discover or understand (**intellexisse**) its obliquity (II.6.31) but there is no explanation of this concept just as there is no precise definition of the ecliptic as the great circle through the middle of the zodiac. The maximum deviations of the planets as given by Pliny are listed in the table.

Table of maximum ´latitudes´

Saturn	$+/- 1°$
Jupiter	$+/- 3°$
Mars	$+/- 2°$
Sun	$+/- 1°$
Venus	$+/- 8°$
Mercury	$+5°$ to $-3°$
Moon	$+/- 6°$

We notice that there is no symmetry in the case of Mercury, and that the sun is credited with a maximum ´latitude´ of one degree. This shows that he is unaware that the path of the sun is the ecliptic circle bisecting the zodiacal belt. On the contrary, Pliny says that the sun travels along a sinuous, serpentine course - **flexuoso draconum meatu** - across the two middle parts of the zodiac. This was a common idea in pre-Hipparchan astronomy although the usual value of the maximum latitude of the sun was half a degree. It may have resulted from the fact that the obliquity of the ecliptic was often supposed to be one fifteenth of a complete circle, or 24 degrees, instead of the more correct value of 23 and a half degrees.(16) It is worth noticing that Pliny seems to be uninterested in the exact positions where the planets have their greatest deviations from the middle of the zodiac,

and in the position of the ´nodes´ where they cross
the middle line. On the whole he makes no attempt
to connect the motion in latitude with the motion in
longitude.

The behaviour of the sun and moon
As we have seen above Pliny is aware of the variation of the length of the longest day with the noon
altitude of the sun at the equinoxes, but unable to
describe this relationship in precise terms. He
has also very little to say about the annual
progress of the sun through the **signifer**, on which
he distinguishes four cardinal points by the two
dates on which day and night are equal (the
equinoctial dates) and the two dates on which the
sun changes its course (the solsticial dates). But
it is worth noticing that he does not place the
corresponding points at the beginnings of the
respective signs. Accordingly the four seasons are
not marked by the entrance of the sun into Aries,
Cancer, Libra and Capricornus. Pliny explicitly
states that the four cardinal points are located at
the eighth degree of their respective signs, so that
for example, spring begins when the sun is a Aries
8°. This norm for the zero point of the ecliptic
ultimately goes back to the so-called System B of
Babylonian astronomy and was abandoned by Hipparchus
among the Greeks, if not earlier. This is another
example of the antiquated concepts which survived in
Pliny, and - to be fair - also in other authors like
Manilius, Geminus and Columella.(17)

Much more detailed is Pliny´s account of the
phenomena of the moon, of which he rightly says that
by her manifold riddles she has tortured the wits of
the observers (II.6.41) among whom Endymion (sic!)
was the first who discovered all the facts about the
behaviour of the moon (II.6.43). Pliny´s own
account is a medley of astronomical facts, physical
speculations and astrological asides. He describes
the phases of the moon in some detail, taking them
as evidence that the moon gets its light from the
sun. The moon is said to complete its orbit in 27
and a third days, after which she is invisible for
two days so that at the 30th day at the latest she
sets out on a new course. This is the closest
Pliny comes to realising the connection between the
sidereal and the synodic months, neither of which
are explicitly mentioned. The invisibility around
the conjunction is explained by the statement that
here the moon comes to within 14 degrees of the sun.

The waxing moon is said to shine for 47 and a half minutes longer and the waning moon shorter by the same period day by day (II.11.58). Both the sun and the moon are nourished by moisture from the earth which dries up quickly in the heat of the sun, but evaporates more slowly in the fainter light of the moon (II.6.46) – an old and familiar idea which, by the way, presupposes that there is no impenetrable barrier between the elementary and lunar spheres.

Pliny makes much of the influence of the moon upon water. Thus he mentions not only that "clever researchers" have shown that it governs the growth of oysters and other shellfish, but also that the tides are due to the combined effects of the moon and sun. Their period is therefore equal to the time between two risings of the moon (II.99.212), which Pliny here takes to be 24 equinoctial hours, which must be used here instead of the unequal hours of everyday life. From what was said above, he ought to have been able to calculate a tidal period of 24 hours, 47 and a half minutes, but of this there is no trace in his theory, which is clearly derived from Posidonius, who is not mentioned by name here but quoted in book I among the authorities for book II. The fact that the tides may sometimes be delayed a couple of hours relative to the calculated time is explained on the assumption that the tidal influence of the sun and moon travels more slowly through space than their optical images (II.99.216).

Another joint effect of the sun and moon is found in the eclipses, in which Pliny is highly interested since they are the most marvellous phenomena in nature and also portents of things to come (II. 6.46). Accordingly he copies records of a great number of eclipses. It is interesting to notice that Pliny´s record of the eclipse that took place before Alexander´s victory at Arbela (September 20th, 331 BC) is more precise than the record of the same eclipse in Ptolemy´s **Geography** (II.72.180; cf. **Geography** I.4). Another interesting record refers to a lunar eclipse taking place at a time when both sun and moon were above the horizon (II.10,57). This may be the eclipse mentioned by Cleomedes and used in evidence of the existence of atmospheric refraction.(18) It is clear that Pliny knows the general explanation of both solar and lunar eclipses (II.7.47), but the details of their geometrical theory escape him. He is correct in stressing that a correct calculation of the shadow of the earth or

moon presupposes that we know the sizes and the distances of both these bodies relative to the earth.
But when he continues by saying that it would be
impossible for the sun to be totally eclipsed from
the earth by the passage of the moon between them if
the earth were larger than the moon (II.8.49) he is
clearly out of his depth. Yet he seems to be aware
of the fact that a solar eclipse is visible only
from a limited part of the earth (II.10.57).

Pliny gives much attention to eclipse records,
stating for instance that it is certain that
eclipses repeat themselves with intervals of 223
months (II.10.56). This is the famous Saros
Period, which is of Babylonian origin but known
already to Hipparchus, who improved upon it.(19)
But there is no evidence that Pliny had any idea why
this is an eclipse period, that is, that 223 synodic
months are approximately equal to 242 draconitic
months, one draconitic month being the time in which
the moon reverts to the same latitude, in particular
to the latitude zero at the nodes of its orbit,
where there is a chance of meeting the sun (or the
shadow of the earth) so that an eclipse may take
place. All this seems to be beyond Pliny, although
it was familiar to astronomers long before his time.
On the other hand he seems to have known that only
the possibility of an eclipse may be predicted by
the Saros Period, but not its actual occurrence,
which can be prevented by other factors. Here
Pliny mentions clouds, but also that "the globe of
the earth may be an obstacle to the convexity of the
world" (II.10.56) - a somewhat cryptic statement
which may be taken to mean that an eclipse may be
invisible because the eclipsed luminary is below the
horizon of the observer.

Since eclipses occur near the nodes of the lunar
orbit, they must take place at, roughly speaking and
ignoring the slow motion of the nodes, opposite
parts of the heavens, so that the usual interval
between eclipses of the same kind must be about six
months. Pliny tells us however that Hipparchus had
found that lunar eclipses may be only five months
apart, and that there may be solar eclipses only one
month apart, although in the latter case they are
not visible from the same place on earth (II.10.57).
There is no reason to disbelieve this bit of
information on Hipparchus, whose work on eclipses
will be mentioned below in another connection.

Planetary Motion

Having dealt with the sun and moon, Pliny is now left with the five remaining planets. They fall into two distinct groups. Saturn, Jupiter and Mars are the superior planets which move in the region above the orbit of the sun. On the other hand Venus and Mercury are called inferior since they are supposed to belong to the region between the sun and the moon. Pliny´s account of the phenomena of the superior planets (II.12, 59-60) can be summarised as follows (cf figure 1):

Elongation	Phenomenon	Characteristics
11 degrees	V1	First visibility (heliacal rising) in the morning
120 degrees	S1	First stationary point where the motion in longitude stops
180 degrees	0	Opposition. The planet rises in the evening. (At sunset)
240 degrees	S2	Second stationary point where the motion in longitude stops once more
348 degrees	V2	Last visibility (heliacal setting) in the evening

The account is not complete since Pliny does not here mention the fact that between S1 and S2 the motion of the planet is retrograde (from east to west) - an oversight which is remedied later in another connection (II.13,69).

The numbers on the table call for some comments. They obviously rest on the simplified assumption that these ´synodic phenomena´ depend only on the elongation of the planet from the sun. This leads to inconsistent values for the first and last visibility which are said to occur when the planet is 11 degrees behind or 12 degrees in front of the sun respectively; furthermore Pliny also maintains that the planets can be seen occasionally at an elongation of only 7 degrees (II.11,58). Clearly

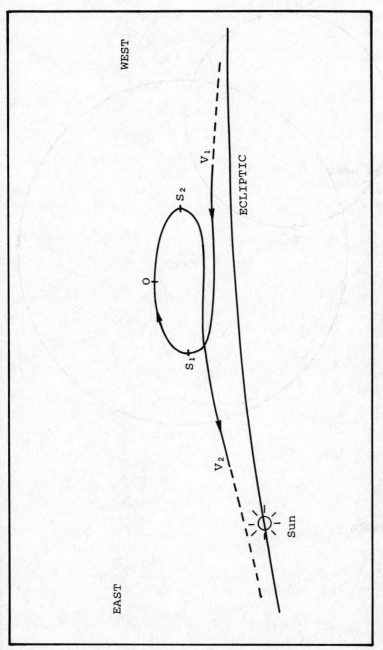

Figure 1 : The synodic phenomena of the superior planets.

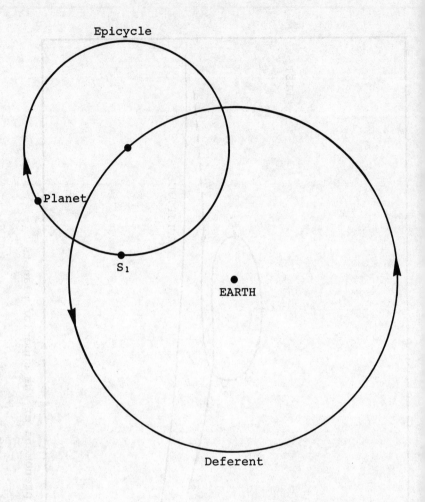

Figure 2

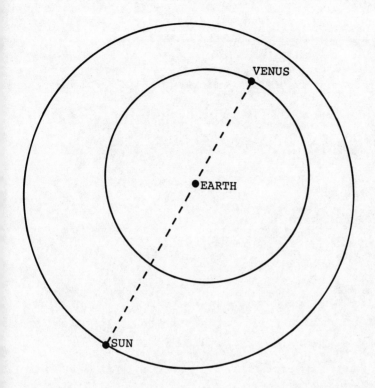

Figure 3

he has no idea that the decisive factor is not the elongation as such, but the difference in altitude between the sun and the planet at positions near the horizon, combined with individual brightness of the planets. The actual periods of invisibility are given as 160 to 170 days for Saturn and Mars, and 26 to 36 days for Jupiter (II.15, 78). There is no attempt to explain why they are not constant.

As for the other numerical values it is obviously correct to place oppositions at elongations of 180 degrees since this is a matter of definition. On the other hand the elongations of 120 degrees and 240 degrees for the first and last stationary points respectively seem to be round numbers. They are no doubt of Babylonian origin(20) and should not be the same for all the planets, a fact which may be reflected in Pliny's assertion that Mars is stationary at elongations of 90 degrees from the sun. Pliny operates with stationary periods, or the time in which a planet remains in the same sign of 30 degrees. Thus Saturn and Jupiter are said to remain stationary within the same sign for four months at a time, while Mars is stationary (in this sense of the word) for six months (II.12,60).

The motion of the inferior planets Venus and Mercury is different in so far as they never separate more than a certain number of degrees from the sun. The maximum elongation is said to be 46 degrees for Venus and 23 degrees for Mercury (II.14, 72). Also here there is a slight inconsistency since the value for Mercury is also given as 22 degrees on the authority of Cidenas and Sosigenes (II.6,39). This does not trouble Pliny who is much more worried by the fact that an inferior planet does not always reach this greatest possible elongation, a fact which is said to have no theoretical explanation - **ratio canonicos fallit** - (II. 14,73). Since the two planets have limited elongations they will move to and fro relative to the position of the sun, and their synodic phenomena of rising and setting become more complicated than those of the superior planets, as Pliny describes in some detail. Here we shall only mention that he assumes that the arc of invisibility is limited by elongations of 11 degrees (or 12?) as in the superior planets, and that the lengths of the periods of invisibility are said to be 52 to 69 days for Venus and 13 to 17 days for Mercury (II.15,78).

Planetary theory
Not satisfied with a mere description of the visible

abandons his role of being a compiler of fact in order to contribute himself to the progress of astronomy with an account which "will differ in many points from that of our predecessors" (II.13,62). Who these predecessors are is not mentioned, although Pliny wishes to give them full credit for having first shown the way to an understanding of these matters. Unfortunately one has to admit that his own attempt to improve upon them is not at all successful, and had better been passed over in silence were it not so that amidst his naive and confused statements it is possible to a certain extent to obtain a vague glimpse of the state of planetary theory in the period between Hipparchus and Ptolemy. Pliny maintains that there are three different causes of the behaviour of the planets and begins as usual with Saturn, Jupiter and Mars.

The first cause of the motion of the superior planets is that they move on circles which the Greeks call **apsides,** a word for which Pliny finds no Latin equivalent and therefore retains in the following. Each planet has its own circle, the centre of which does not coincide with the common centre of the earth and the universe. To define such a circle it is necessary to indicate its radius and the position of its centre, but all we can find in Pliny is an indication of the direction in which the centre is to be found or, in modern terms, the direction of the apsidal line from the earth to the ´highest point of the circle. These values are listed in the following table (in column 1) from which it is seen that the ´apogee´ is always placed in the middle of the respective sign (II.13,64).

Table of apogees

	I		II	
Saturn	Scorpius	15°	Libra	20°
Jupiter	Virgo	15°	Cancer	15°
Mars	Leo	15°	Capricornus	28°
Sun	Gemini	15°	Aries	29°
Venus	Sagittarius	15°	Pisces	27°
Mercury	Capricornus	15°	Virgo	15°
Moon	Taurus	15°	Taurus	4°

This shows that Pliny cannot claim Hipparchus as one of his predecessors, since Hipparchus had shown that the apogee of the sun must be placed at Gemini five and a half degrees. The stereotype values of the Plinian apogees show that here he connects with an older tradition, confusing the apogees with a norm placing the cardinal points of the ecliptic in the middle of their signs.(21)

That Pliny did not quite understand the geometry of his eccentric circles becomes evident when he goes on to the second cause of planetary motion, that is, that the planets have other **apsides** than those listed in column I (II.13,65). This second set is listed in column II. They do not agree with the first set, neither with respect to the sign nor to the degree within the sign. Now, a circle cannot have more than one apogee, and this second set of numbers seems rather mysterious until one realises that here Pliny has got hold of the positions of the astrological **exaltationes** of the planets, that is, the places where they exert the strongest influence on things here below. But Pliny seems to be completely unaware of the fact that these exaltations have nothing whatever to do with the motions of the planets. That he nevertheless lists them among the causes of the visible phenomena reveals more than anything else that he did not really know what he was speaking about.

Finally, Pliny has a brief remark on the third cause of the ´altitude´ of a planet (II.13,65) which is not related to its circle, but to the **mensura coeli**, since the eye observes them as rising or sinking through the depth of the air. It is difficult to see what this means, unless Pliny here interprets the reduced brightness of a star near the horizon as caused by greater distance.

Everything considered, Pliny´s ´theory´ of the superior planets seems so confused and unintelligible that one is tempted to dismiss it as so much nonsense. Nevertheless, it can be taken as a testimony that planetary theory at his time or even before tried to deal with the superior planets in the same way as Hipparchus had dealt with the motion of the sun, using a model with an eccentric circle. That Pliny was no mathematician explains that he left all the details of such models out of consideration, confining himself to a very few obvious consequences of the motion on an eccentric. Thus he is aware that the motion of the planet will seem to be slower when it is on the ´upper´ than when it is on the ´lower´ part of its circle (II.13,64). This is true on the assumption that the motion of a circle is uniform with respect to the centre; but this fundamental presupposition of Greek mathematical astronomy is not mentioned at all. The best one can say is that Pliny had, perhaps, a vague

idea of the so called ´first anomaly´ of planetary
motion, that is, the variation of the apparant
velocity of a planet during a complete revolution.
   The question is if Pliny also knew that the
´second anomaly´ or the retrograde motion between
the two stationary points cannot be explained by a
model with only one eccentric circle.   The answer
depends on the interpretation of a few passages
(II.13,69-70) in which Pliny says, first, "that
there is no doubt that the three superior planets
increase their motion in their morning risings and
diminish it from the first to the second stations".
This is correct if ´motion´ here means the motion in
longitude, since the planet is here moving eastwards
from V2 to S1 (see Figure 1) and westwards from S1
to S2.   Second, Pliny continues by saying that in
"their first stations their altitude also is
increased since then the numbers first begin to
decrease and the stars to recede" (II.13,69).   This
seems to indicate that Pliny is aware that the path
of a superior planet has the form of a loop, as
shown in Figure 1.   This might perhaps be taken as
evidence that some kind of epicyclic model is behind
this conception.   In such a model the eccentric
deferent does not carry the planet directly, but
carries the centre of a circle called the epicycle
on which the planet rotates.   It is seen from
Figure 2 that if the planet is lifted upwards at the
first station S1, as Pliny seems to say, then the
motion on the epicycle must have the opposite
direction of the motion of the epicycle on the
eccentric.   Models of this kind are known from a
couple of astronomical papyri and Aaboe has examined
the conditions under which they will work.(22)
   Pliny also attempts to give a physical explanation of the increase in the planet´s distance at
the first station.   This is caused by the rays of
the sun which prevent the planets from going on when
the elongation is 120 degrees and "lift them up by
their fiery force" (II.13,70).   This means that
Pliny assumes that the solar rays exert a mechanical
force on the planets which is strong enough to stop
their motion in longitude when the distance between
the sun and the planet is diminished to 120 degrees.
This so called ´heliodynamic theory´ seems to be
Pliny´s own idea.   It is not a very fortunate explanation of what happens around the first station
since the planet is not here running up towards the
sun.   On the contrary, the sun is here receding
from it so that its alleged mechanical action would
diminish, at least according to ordinary views on

the nature of forces. We shall not here pursue the
´heliodynamic´ hypothesis further, since it was a
vague notion which made no impact on Greek astronomy
in general.

Turning from the superior to the inferior
planets, Pliny asserts that their theory "is more
difficult and has been explained by nobody before
ourselves"(II.13,71). This points to the existence
of an already established theory of the superior
planets, perhaps of the epicyclic variety outlined
above. It also seems to indicate that no such
theory existed for Venus and Mercury, or at least
that Pliny did not know about it and tried to work
out his own explanation. The result is an attempt
to explain why the inner planets have limited
elongations and are never in opposition to the sun.
The words of Pliny are here extremely vague and
have never found any satisfactory explanation.(23)
Thus he says of Venus and the sun that **non
possunt abesse amplius** (than their maximum elongation
of 46 degrees) **quoniam curvatura apsidum ibi non
habet longitudinem maiorem** (II.14,72). Now if both
Venus and the sun are supposed to have eccentric
apsides it is clearly wrong that they cannot be in
opposition, as seen from a glance at Figure 3.
Vogt has tried to save the wording of the text by
assuming that **longitudo** here might mean not the
angular but the linear distance of the two bodies;
this would explain that the reason for the
limitation of this distance is said to be the
curvatura of the orbits.(24) On the other hand it
is difficult to believe that Pliny speaks of
anything else than the angular distance in terms of
which he describes the visible phenomena of these
planets.

Cosmic dimensions
Here and there in the **Natural History** Pliny refers
to cosmic dimensions although he considers it to be
sheer madness to try to find the size of the universe as a whole (II.1,3). All we can hope for is
to find the dimensions of some of its constituent
parts. Thus the size of the earth is dealt with in
the very last chapter of book II where Eratosthenes
is mentioned as the great authority on the subject.
Pliny correctly quotes his value of the circumference of the earth as 252,000 **stades**, which he
translates into 31,500 Roman miles (II.112,247).
The measurement is called an audacious venture, but
it is based on such subtle reasoning that one ought
to be ashamed of not believing in the result. Pliny

adds that Hipparchus had added a little less than 26,000 **stades** to the value of Eratosthenes - an assertion that cannot be verified from other sources. It is worth noting that Pliny does not refer to the value of 240,000 **stades** proposed by Posidonius, but that instead he tells us - as an example of the **vanitas Graecae** - that when the geometer Dionysiodorus of Melos died, a letter was found in his tomb stating that the distance to the bottom of the earth was 42,000 **stades**. From this some geometers inferred that the circumference must be 252,000 **stades** (II.112,247). This calculation implies that **pi** = 3, and Pliny knows that this is wrong since he states that the diameter of a circle is **tertiam partem ambitus et tertiae paulo minus septimam** (II.21,86), that is, a little less than 1/3 + 1/21 of the circumference. This shows that the text must be corrupt since he immediately afterwards gives 7/22 as the correct value (II.21,87). In the last paragraph Pliny cryptically adds that one should add 12,000 **stades** since this would make the earth 1/96 part of the whole world. It is difficult to see what this means. But if the radius of the earth is 42,000 **stades**, (according to the legend of Dionysiodorus) the geometers ought to have found a circumference of 42,000 · 22/7·2 = 264,000 **stades**, or 12,000 **stades** more than the Eratosthenes value. Perhaps this is meant as a criticism of Hipparchus.

As for the distances within the universe, Pliny quotes Posidonius for the statement that there are 2,000,000 **stades** from the cloudy atmosphere to the moon and a further distance to the sun of 500,000,000 **stades** (II.21,85). On the other hand, Pliny says that many people have assumed the sun to be 19 times as far away as the moon (II.19,83). This looks like a reminiscence of Aristarchus´ result, that is, that the distance of the sun is between 18 and 20 times the distance of the moon.(25) Finally he mentions that Pythagoras found the distance of the moon from the earth to be 126,000 **stades**, the distance of the sun from the moon to be twice this value, and the distance of the zodiac (that is, the fixed stars) from the sun to be three times this number (II.19,83), a view which is said to have been held also by the Roman astronomer Sulpicius Gallus.

Besides these numerological speculations Pliny also refers to another tradition according to which the orbit of the sun is said to comprise almost 366 **partes,** and that "accordingly", the distance of the sun is about 1/6 of the circumference (II.21,86). This has, of course, no astronomical significance. When he continues by saying that the distance of the moon is 1/12 of this orbit (the sun) he implies that the sun is only twice as far away as the moon, in disagreement with all the values previously quoted. Finally Pliny refers to an Egyptian theory by Petosiris and Nechepsos according to which one degree of the lunar orbit equals 33,000 **stades**, that of Saturn twice as much, and that of the sun the mean of these two values; but this is discarded as a shameful calculation (II.21,88).

Thus Pliny presents his reader with a variety of mutually inconsistent values of cosmic dimensions without stating his own preference for any of them. Clearly this subject was of no great importance to him.

Pliny and the History of Science

The conclusion to be drawn from the preceding sketch of Pliny's astronomy must be that he was no astronomer, but a rather incompetent compilator of astronomical lore culled from a variety of sources, some of which were not of the purest water. Thus it is impossible to give him any place at all in the development of astronomy. On the other hand, his rich collection of detailed information, good or bad as it may be, raises the question whether we should not regard him more as a historian of astronomy than as an astronomer in his own right.

There is no doubt that in general Pliny thought highly of science as an essential factor of civilization, and that he deplored the prevailing ignorance about it. Thus with regard to the phenomena of the moon he writes that

> we are surely not grateful to those whose labour and care have provided us with insight into this luminary, the human mind being so strange that we delight in filling our annals with bloodshed and slaughter, so that the crimes of man may be known to those who are ignorant of the universe (II.16,43).

This sympathetic evaluation of the history of science over and above the traditional political history of wars and battles is the background of the potted history of science and technology which Pliny included in book VII of the **Natural History**. Here he gives a list of people who have gained distinction in science and therefore ought to be reckoned

among the flowers of mankind (VII.37,123). Among
these worthies he reckons Berosus the Chaldean,
whose marvellous astronomical predictions caused the
Athenians to offer him a statue with a gilt tongue;
Archimedes, whose knowledge of geometry and mech-
anics was so remarkable that the general Marcellus
forbade (in vain) his troops to injure him at the
capture of Syracuse; Chersiphron the architect of
the temple of Diana at Ephesus; Philo, who made a
dockyard for 400 ships at Ahens; Ctesibus, who discovered the theory of pneumatics and imvented
hydraulic apparatus; and Dinochares, who acted as
surveyor for Alexander at the foundation of Alexandria. These are excellent examples and more
convincing than the following in which Pliny strives
hard to convince us that the Romans surpassed the
Greeks in almost every field.

Particularly amusing is Pliny´s account of the
gradual development of a public time service in Rome
– a field in which the city played no conspicuous
role. On the contrary, Pliny admits that of the
three things generally acknowledged by all - that
is, the use of the alphabet, the practice of shaving
and the division of the day into hours - this latter
usage came to Rome only at a late date. The **Twelve
Tables** ordered only sunrise and sunset to be
observed; a few years later also the time of noon
was daily proclaimed by a consular official when he
saw, from the Senate House, the sun standing between
the **Rostra** and the **Graecostasim.** When the sun
sloped down from the Maenian column towards the
prison he also announced that the last hour of
daylight was at hand - but only on clear days...
(VII.60,212).

This primitive time service (based on a rough
estimate of the azimuth of the sun) lasted until the
time of the first Punic war, when a sundial was
brought as a trophy from Catania and set up on a
column in Rome. Although the lines of this dial
did not agree with the hours (it was constructed for
a lower latitude) the Romans used it for 99 years
until Quintus Marcius Philippus presented the town
with a properly designed dial (VII.60,214) of which
they complained, nevertheless, that it was of no use
in cloudy weather. Then, at last, Scipio Naso took
the matter in hand and in 595 AUC (159 BC) erected a
waterclock in a roofed building (VII.60,215). This
horologium showed the hours both by day and by night
regardless of the weather, and taught the Roman
people to observe the division of time into hours,
no doubt with the Tower of the Winds in Athens as a

model.

It is passages like these which here and there make the **Natural History** such delightful reading. There is no essential reason to doubt the veracity of the account, in particular since it does not depict the Romans in too flattering a light. However, in other cases one has to be more circumspect. Thus it is best to discard the statement that, according to Epigenes, the Babylonians possessed bricks inscribed with astronomical records for the past 730,000 years, a number which Berosus and Critodemus more modestly put at 490,000 years (VII. 56,193), and this is all the more easy since Pliny seems to have doubted these exaggerated fantasies himself. On the other hand there are a number of passages in which Pliny provides information which is not found in other authors and which - if it be true - is of the greatest interest to the history of ancient astronomy. We shall briefly discuss two such passages.(26)

Two Hipparchan problems
The first example is Pliny´s famous account of Hipparchus´ discovery of a new star. The text runs as follows:

> Idem Hipparchus numquam satis laudatus ... nouam stellam (et aliam) in aeuo suo genitam deprehendit eiusque motu qua (die) fulsit ad dubitationem est adductus, anne hoc saepius fieret mouerenturque et eae, quas putamus adfixas, ideoque ausus rem etiam deo inprobam, adnumare posteris stellas ac sidera ad nomen expungere, organis excogitatis, per quae singularum loca atque magnitudines signaret, ut facile discerni posset ex eo non modo an obirent ac nascerentur, sed an omnino aliquae transirent mouerenturque, item an crescerent minuerenturque, caelo in hereditate cunctis relicto, si quisquam, quo cretionem caperet, inventus esset (II.24,95).

This passage has always given rise to great difficulties. It pretends that during his lifetime Hipparchus detected a new star and consequently began to doubt whether this was a unique event, or whether it might possibly happen again. In order to make it possible to decide this question he made the audacious resolution to survey the heavens and leave to posterity a list or catalogue of the stars.

We shall not here go into the tangled question of Hipparchus´ catalogue of stars,(27) but only briefly touch on the reason why so many commentators have doubted the historical truth of this account.(28) The text says that Hipparchus detected a new star, but was led into doubt (of its being a fixed star) by its **motus qua fulsit.** It is clear that if **motus** means movement from one position to another, the phenomenon cannot have been a fixed star. In this case it may have been a comet or meteor, and it is worth noticing that Pliny tells us of comets just before, and of meteors just after, the passage in question. However, this objection has no real weight since the words **Idem Hipparchus** clearly reveal that the whole passage is misplaced, Hipparchus having not been mentioned in the preceding pages (in fact not since II.10,57). Furthermore, a number of manuscripts insert the words **et alia** after **stellam**, perhaps in order to underline that the phenomenon was not a comet like those just described before.

Another difficulty is caused by the words **qua fulsit.** Here **qua** cannot relate to **motus** because of the difference of gender. Several editors have accordingly construed it adverbially, reading - in agreement with some of the manuscripts - **qua die fulsit,** which does not seem very helpful.

To solve the question it should be remembered that **motus** may well be a translation of the Greek **kinesis** which in Greek philosophy was used not only of spatial motion, but of qualitative change as well.(29) If this is so, it is possible so to construe the phrase that Hipparchus was led into doubt concerning the star because of the change (that is, of brightness) by which it flared up, and that in consequence he began to record both **loca atque magnitudines** (positions and brightnesses) of the fixed stars. This does not prove, of course, that Hipparchus really discovered a **nova.** All we can say is that Pliny´s text does not exclude this possibility.

Also another bit of information on Hipparchus is known only from Pliny, who maintains that the correct explanation of eclipses was first given by Thales among the Greeks and that it became known to the Romans through Sulpicius Gallus. He then continues by saying that

Post eos utriusque sideris in sexcentos annos praececinit Hipparchus, menses gentium diesque et horas ac situs locorum et uisas populorum

> complexus, aeuo teste haud alio modo quam
> consiliorum naturae particeps.(II.9,53)

Taken at its face value this passage pretends that
Hipparchus **praececinit** - that is, prophesied, fore-
told or predicted - the course of both the sun and
the moon for a period of 600 years, taking also
geographical factors into account, if that is what
the obscure phrase **menses gentium...complexus** means;
if so, it agrees with the statement by Achilles
Tatius that Hipparchus wrote a treatise on solar
eclipses in each of the seven climates.(30)

   Now, all commentators have refused to believe
that Hipparchus calculated eclipses for no less than
600 future years - a period in which there are
almost 15,000 possible eclipse situations (new and
full moons). This would have been an overwhelming
work of tedious and difficult numerical calculations
which - if it were ever performed - would surely not
have disappeared without any trace whatever in Greek
astronomical literature. A possible explanation of
what might be behind Pliny´s words has been proposed
already by Lalande who said "Hipparque fit un
recueil des éclipses de soleil et de lune observées
des Chaldéens".(31) This idea has been taken up by
Professor Neugebauer, who draws attention to the
fact that Hipparchus was interested in the time
intervals between successive eclipses and therefore
would have studied earlier eclipse records. Now,
the first known ´Chaldean´ (that is, Babylonian)
record of an eclipse (of the moon) dates from 721
BC, while the last eclipse observed by Hipparchus
himself was in 127 BC. This time interval is very
nearly 600 years. In consequence it is a plausible
hypothesis that Hipparchus did not predict eclipses
for 600 years into the future, but that he listed
eclipses for 600 years into the past.(32)

   In consequence, there is only little doubt that
Pliny referred to a sensible project actually
carried out by Hipparchus. The question is, what
went wrong with his account? The only stumbling
block is the word **praececinit**, which is derived from
praecano, the original meaning of which is to ´sing´
of something before it happens. But since Hipparchus told of eclipses after they happened the
prefix **prae-** is clearly impossible. It would be
better to read **recicinit**, since **recano** means to
´sing´ of something after the event, in particular
to undo the effect of a spell or charm by a magical
incantation. Here again it is worth noticing that
at least in one other place in the **Natural History**

193

the verb **praecanere** is used in a sense referring to something in the past, that is, to undo the harm of the bite of a viper (XXIX.21,69) and that at least one editor has not hesitated to emend this word to **recanere**. If this substitution of a **prae-** by a **re-** is acceptable we arrive at the same result as before, namely that Pliny´s text makes sense also in this case and provides us with a piece of genuine information on the astronomical activity of Hipparchus.

NOTES

1.   For the Latin text of the **Natural History** we have usually relied upon the edition of J Beaujeu, **Pline l´Ancien: Histoire Naturelle, Livre II**, Paris, 1950 (Collection G Budé). Many of the English quotations are adapted from the Loeb edition by H Rackham, **Pliny: Natural History**, vol. 1, London, 1958 (1938) which must be used with some care.

2.   These quotations are from O Neugebauer, **A History of Ancient Mathematical Astronomy**, Berlin, Heidelberg, New York, 1975, p. 802, a standard work to which we shall refer as **HAMA**; on Pliny in general see H le Bonniec, **Bibliographie de l´Histoire Naturelle de Pline l´Ancien**, Paris, 1946 (Collection d´Études Latines, Série sci. 21); W Kroll, "Plinius"(5) in Pauly-Wissowa 21(1) (1951) pp. 271-439, and R König and G Winkler, **Plinius der Ältere, Leben und Werk eines antiken Naturforschers**, Munich, 1979.

3.   These figures seem to be underestimated by Pliny. A recent survey has shown that he refers to 473 authors and mentions 34,707 "remedies, researches and observations"; see D E Eichholz, "Pliny", **Dictionary of Scientific Biography**, XI (1975) pp.38-40. On Pliny´s authorities see F Münzer, **Beitrage zur Quellenkritik des Plinius**, Berlin, 1879.

4.   Cicero, **De Divinatione**, II.43,90, on the twins Procles and Eurysthenes who both became kings of Sparta but did not live the same number of years.

5.   Augustine, **Confessions**, V.6 and, in more detail, **De Civitate Dei**, V.1-6, where the Biblical twins Esau and Jacob are used as an example frequently quoted against astrology during the following centuries.

6.   Here the text is uncertain. Rackham has **Quisquis est Deus, si modo est aliquis,** translating this as "whoever God is - provided there is a God"

(Loeb ed., 178-9). Beaujeu (following Detlefsen) reads **Quisquis est deus, si modo est alius** (Budé ed.,12) which seems to be the better reading since Pliny speaks here of the attributes which make God different from anything else.

7. See, e.g. O Pedersen, "The ecclesiastical calendar", in G V Coyne, M A Hoskin and O Pedersen, eds., **Gregorian Reform of the Calendar**, Città del Vaticano, 1983, pp. 17-74.

8. On the history of the Greek parapegmata in general, see **HAMA** pp.587ff. and A Rehm, **Parapegmastudien**, 1941 (Abh. Baeyr. Akad. Wiss., Philos. -hist. Abt. NF 19).

9. On Pliny´s vocabulary see D J Campbell, **C. Plinii Secundi Naturalis Historiae Liber Secundus**, Aberdeen, 1936 (not too helpful with respect to astronomical terminology).

10. See Geminus, **Isagoge**, ed. G Aujac, **Geminos, Introduction aux Phénomènes**, Paris, 1975, pp.98-108 (Coll. G. Budé), and H Vogt, **Der Kalendar des Claudius Ptolomaeus**, 1920 (Sitz.- Ber. Heidelberg Akad. Wiss., Philos.-hist, Kl., no. 15).

11. See J E Skydsgaard, **Varro the Scholar**, Hafniae, 1968, pp. 43-63

12. **HAMA** p. 612

13. See E Honigmann, **Die sieben Klimata**, Heidelberg, 1929, in particular pp. 31-54.

14. **HAMA** pp. 729ff. and 747ff.

15. **HAMA** p. 286

16. **HAMA** pp.733ff.

17. **HAMA** pp. 594ff.

18. Cleomedes, **De Motu Circulari Corporum Celestium**, ed. H Ziegler, Leipzig, 1891, II, 1, pp. 122ff.

19. On the Saros period see **HAMA** p. 310.

20. **HAMA** p.411

21. On the variety of such norms in antiquity see **HAMA** pp. 594-600.

22. A Aaboe, "On a Greek qualitative planetary model of the epicyclic variety", **Centaurus**, 9, (1963) pp, 1-10.

23. "An unintelligible jargon, calculated to impress his readers but not worthy of our attention here", W H Stahl, **Roman Science**, Madison, 1962, p. 11; "Alles das entzieht sich der Verstandnis, weil Plinius es vermieden hat, die mathematischen Grundlagen zu geben", W Kroll, **De Kosmologie des Plinius**, Breslau, 1930, p.21.

24. H Vogt, in W Kroll, op.cit., note 23, p. 69f.

25. **HAMA** p. 660

26. These will be dealt with in greater detail in a forthcoming paper.

27. On Hipparchus´ ´Catalogue´ of stars see **HAMA** pp. 285-288, with reference to the relevant literature by Maass, Rehm, Boll and others.

28. See, e.g. J K Fotheringham, "The new star of Hipparchus", **Monthly Notices Roy. Astron. Soc.**, 79 (1919) pp. 162-167, and D H Clark and F R Stephenson, **The Historical Supernovae**, Oxford, 1977, p. 14; cf. **HAMA** pp. 284f.

29. See, e.g. Aristotle, **Physics** III.1 201a 11 and V.1, 224b 11; cf. **Metaphysics** IV.14, 1020b 20 and IX.6, 1048b 16ff.

30. According to G J Toomer, "Hipparchus", **Dictionary of Scientific Biography**, XV, (1978) p. 216.

31. Lalande, **Astronomie I**, para. 322, 3rd ed., Paris, 1792, p. 113

32. **HAMA** p. 319 and p. 366

Chapter Eleven

PLINIAN ASTRONOMY IN THE MIDDLE AGES AND RENAISSANCE

B S Eastwood

Plinian astronomy in the middle ages and renaissance
is a subject as much in need of definition as it is
in need of either exploration or synthesis. What I
propose to offer here is a working hypothesis which
may become a suitable definition of the core of
Plinian astronomy in medieval and renaissance times
when further study has been done. Guided by my
hypothesis, I hope to show the varying historical
vitality of Plinian astronomy. However, I have no
intention of suggesting any final synthesis.

1. **Pliny**
To begin with the **Natural History** itself, we can
find a central theme rather quickly in the first 88
sections of book II, which contains the essence of
Pliny´s astronomy(1) - I call it "his" for the sake
of convenience. Generally Stoic and influenced by
astrology(2), his cosmic system is one of forces,
perhaps tonic forces(3), maintaining order and the
regular patterns of celestial events with respect
primarily to the sun.(4) Schiapparelli´s editor used
the wonderfully suggestive term "heliodynamic"(5) to
describe this system, but the term may be too
suggestive for post-Copernican observers. Nevertheless, the solar focus is the one I want to start
with, for it locates my discussion within planetary
theory, if we can use that word with Pliny. The
Plinian astronomy which I find most noteworthy in
later times is that which deals with (a) the effects
of solar rays, especially planetary stations and
retrogradations, (b) planetary apsides and (c)
planetary latitudes. Commonly included but less
distinctive considerations are Pliny´s descriptions
of planetary order and planetary intervals. All
these elements are fully discussed in book II and it

is the history of book II as well as of these
contents to which we now turn.

Martianus Capella

For medieval Latin authors prior to the mid-twelfth
century introduction of Arabic and Greek astronomy
to the West, three late ancient books were widely
consulted for astronomical information. The works
of Martianus Capella, Macrobius and Calcidius, all
written just beyond the fourth century, were mined
repeatedly from the ninth to twelfth centuries for
knowledge about the heavens, but only the first,
Martianus´ **Marriage of Philology and Mercury** shows
noteworthy commitment to ideas and data about the
planets found in Pliny. In book VIII, a summary of
basic descriptive astronomy,(6) Martianus devotes
rather less than half (sections 850-887) to the
planets. After using Geminus for much of his
material on the planets as well as the constellations,(7) Martianus takes most of his closing
account (sections 879-887), dealing with those
planets other than the sun and moon, from Pliny,
modified somewhat by Calcidius and Theon. Twice
Martianus refers to the percussive and attractive
force of the sun´s rays as the cause of planetary
anomalies, that is, the stations and retrogressions
of the superior planets as well as their resumption
of forward progress(8). For the inferior planets,
Mercury and Venus, Martianus has a special pattern,
circles around the sun, to explain their
anomalies(9). While he almost certainly draws this
pattern from Theon of Smyrna(10), there is room to
speculate that Martianus understands the power of
solar rays to be responsible for holding the two
planets near the sun, thus making the solar ray
theory of Pliny responsible for the anomalies of all
five planets.(11) Martianus adopts Pliny´s fixed
locations for planetary **apsides altissimae**(12),
positions where the planets are at their greatest
distance from the earth. However, only the outer
three planets can be assigned such apsides in
Martianus´ astronomy, for the circumsolar paths of
Mercury and Venus result in their far-points
appearing at various places in the zodiac, not at
Pliny´s fixed points. So Martianus gives the
apogees as Mars in Leo, Jupiter in Virgo, Saturn in
Scorpius(13). Further connection to Pliny appears
in the Capellan adoption of Pliny´s astrological
apsides, or exaltations, for the three outer
planets, that is, Mars at Capricorn 28, Jupiter at

Cancer 16, and Saturn at Libra 20(14). The detailed description of Venus is mostly from Pliny(15). With a strong Plinian thread, Martianus Capella´s textbook on astronomy was an essential basis for scientific knowledge in the ninth to twelfth centuries and continued to be copied and referred to thereafter(16).

Isidore and Bede

In the early seventh century we find Isidore of Seville´s **Liber Rotarum** (about 613) using Pliny very little for astronomy and only for phrases and names. At one point Isidore refers the effects of solar rays on planetary motion - retrogrades and stations - but draws this from the poet Lucan.(17) In the nineteen chapters devoted to astronomy and cosmology, Isidore uses Pliny only a few times and pays much more attention to Hyginus and the church fathers.(18) In the whole book Isidore shows knowledge or verbatim use of Pliny at eighteen points, especially concerning weather and winds.(19) A work attributed to Isidore, composed after his **Liber Rotarum**, used Pliny´s descriptions of the planets´ appearances, their orbital periods, and the planetary apsides. The work is the so-called **Scholia Sangermanensia** to the **Aratea**, actually a modification of the seventh-century **Aratus Latinus** translation.(20) Accepting the attribution to Isidore, we still find him preferring Hyginus and other sources for most of the planetary material in the **Sangermanensia.**(21)

A sharp contrast to Isidore´s **Liber Rotarum** is made by Bede´s **De Natura Rerum** (about 703) which, according to a contemporary, he made in order to prevent his students from reading what he called the "lies" in Isidore.(22) Bede´s opinion of Isidore was not high and, among other differences, the English scholar leaned much more than his predecessor upon Pliny for astronomical doctrine. Of 51 chapters Bede derived much or most of his material for 32 from Pliny, 13 from Isidore.(23) Furthermore, at only two places, chapter 47 (on eight latitudinal circles for varying day lengths) and chapter 51 (on the tripartite division of the earth) are there Plinian materials not found in book II of the **Natural History** and not available in other sources.(24) Bede draws much of his meteorology, his doctrines on the moon and on eclipses, a great deal of astronomical nomenclature, and all of his planetary descriptions and theory from Pliny.

In six consecutive chapters (11-16) Bede sets out
his planetary doctrine.   In the latter half of
chapter 11 Bede uses Pliny´s statement that solar
rays influence the stars and can thereby bring
changes in the weather.(25)   Chapter 12 simply
describes the general contrary motions of stellar
caelum and the seven wanderers and the fact that
they are variably inclined to the plane of the
zodiac.   The chapter closes with the assertion
"Nevertheless, hindrances and anomalies, whether
retrograde or stationary, are made (to the planets)
by the rays of the sun".(26)   Bede takes this
phrasing from Isidore, who got it from Lucan.(27)
It is interesting to find Bede inserting this
Isidorean phrasing of the heliodynamic idea into a
chapter which is drawn mostly from Pliny.   Chapter
13 gives the Plinian/Ptlolemaic order of the
planets, their periods, and the lengths of intervals
of invisibility in the same words as Pliny.(28)
We can see that Bede still has available in the
early eighth century either long running sections of
book II or extensive excerpts which are worded and
phrased as Pliny gave them: witness chapters 14 to
16.   The first of these defines and lists the
planetary **apsides altissimae**, or apogees, closing
the chapter with a direct reference to the **Natural
History**, "concerning which matters, if you wish to
know more, read Plinius Secundus, from whom we have
excerpted...".(29)   One of the things that makes
the quotations from Pliny interesting is Bede´s
careful omission of the astrological apsides, while
using the material before and after this section in
Pliny´s text.(30)
Chapter 15 abstracts verbatim from Pliny
regarding the planets´ colours, which change either
towards dullness or greater heat because of the
influences of the stars when they are approached.
The planets are obscured by the sun, apsidal nodes
and the more extreme distances of their orbits.(31)
Possible astrological uses of planetary colours
apparently escaped Bede´s attention.
Bede´s chapter 16 comes directly from Pliny´s
account of the zodiac and the planetary wanderings
across its twelve degrees.   In fact, Venus is so
extravagant as to exceed the zodiacal band by two
degrees, that is, one degree on either side.   The
moon is more orderly and covers exactly the latitude
of the band.   The other planets also follow Plinian
descriptions: Mercury with a latitude of eight
degrees, Mars four, Jupiter three, Saturn two and
and sun two.   The last is surprising, but Bede does

question Pliny in attributing to the sun a serpentine path through the two middle degrees of the zodiac.(32)

Availability in the eighth century

Bede´s use of Pliny has been the focus of disagreement regarding the availability of the **Natural History** in Britain in the eighth century, which is another way of asking whether or not manuscripts of much or all of the **Natural History** entered the Carolingian empire primarily from insular sources. This question is complicated by the clear availability of intermediates, like Isidore of Seville,(33) whose quotations from Pliny can easily set out false scents. For example, Bede cannot be shown to have used directly book XVIII of Pliny, despite superficial similarities of parts of that book with **De Natura Rerum,** 26.(34) It is exactly this book XVIII whose apparent absence in Bede´s Britain makes an English origin unlikely for excerpts from Pliny appearing at the beginning of the ninth century in Munich (CLM 210) and Vienna (lat. 3307) manuscripts of a three book computus and in Vatican (Regin. lat 309) and Madrid (lat. 3307) manuscripts of a seven book computus.(35) E A Lowe´s identification of three palimpsests from the fifth century with texts of Pliny assigns these early materials all (tentatively) an Italian origin.(36) As for the amount of Pliny´s text available to Bede, only book II seems rather undeniably present at Wearmouth-Jarrow, for the material used from books III and VI(37) is so limited that it allows the assumption of longer excerpts or segments of those books rather than a continuous text from book II to book VI.(38)

The Carolingian manuscripts

Beyond Bede we must wait for Alcuin on the continent to carry the story further. Since we do not know what of Pliny Alcuin actually knew, since the three book and seven book **computi** were composed on the continent at the beginning of the ninth century, and since the continental interest in Plinian materials during the ninth century has not yet been traced to insular manuscripts,(39) we can conclude in all likelihood that an influential set of six Plinian astronomical-cosmological excerpts from books II and XVIII(40) was compiled on the continent. These excerpts may have been made before the early ninth century, although we have no strong evidence for that assertion at present. From these excerpts, those which concern astronomy are four: (i) on the order and paths of the seven planets, (ii) on their intervals, (iii) on their apsides and (iv) on their paths through the zodiac.(41)

Vernon King reported on thirty eight manuscripts of these Plinian astronomical excerpts from the ninth to the twelfth centuries in addition to the seven manuscripts from the ninth century for the two **computi.** Compared to the few manuscripts of all or large sections of the **Natural History,** the astronomical excerpts were by far the best known and most widely used portions of Pliny up to the twelfth century. Aside from the thirty folia of long fragments in an eighth century Leiden(42) manuscript, we have seven long manuscripts from the period:

1. s.IX(I) - New York, Pierpont Morgan M.871, bks. I-XVII
2. s.IX - Paris BN 6795, bks. I-XXXVII
3. s.X (or XI or XII) - Florence, Bibl. Riccardiana 488, bks. 1-XXXVI, many lacunae
4. s.XI - Leiden Univ. Bibl. Lipsius 7, bks. I-XXXVII
5. s.XI - Vat.lat 3861, bks. II, 187 -VI, lacunae, a source of Leiden Lipsius 7
6. s.XI(2) - Berlin Hamilton 517, bks. I-XXXI
7. s.XI-XII - Vienna NB 9-10, bks. II-XIX, XX-XXXVII, continuous, by the same hand.

From the twelfth century there are four more codices of half or more of the **Natural History** and one of book II alone.(43) Even with the assumption that other long manuscripts from the ninth to twelfth centuries have perished, the relative survival rate still indicates the predominance of the excerpts in that period.

The astronomical excerpts
Turning to the contents of the astronomical excerpts, we find that each provides the data indicated in its title, and the latter two, on apsides and latitudes, include a clear statement of the power of the sun´s rays to cause planetary anomalies. In connection with apsides the text says at one point that the stations occur when a solar ray strikes the planet at trine and the fiery force drives the planet directly away from us so that it appears not to be moving at all. The violence of the solar ray

then increases and sends the planet into retrograde, especially if it is at its evening rising, to the extremes of its **absis**, where it is farthest from us, appears very small, and also is reduced to minimum angular motion.(44)   The zodiacal sign for each planetary apogee (**absis altissima**) is then given, each said to be in the very middle of the zodiacal sign, with perigee always in the opposite sign. (45)

In the excerpt for zodiacal latitudes of the planets the text says that in a descent after evening rising the solar ray strikes the planet from the other side and brings both a drop in latitide and an altitudinal decline towards the earth.   The effects of solar rays reverse themselves when they strike first one side and then the other of the planet.(46)   The excerpt also presents Pliny´s values for the latitudes and discusses planetary risings and settings in both morning and evening. Finally, periods of planetary invisibility are listed and attributed to combinations of planetary station, apsidal node, and the occultation by the sun´s brilliance.(47)

Diagrams for the excerpts
The four astronomical excerpts appear immediately in the company of illustrative diagrams, and this is especially interesting because **no** longer or complete Pliny manuscript from **any** period contains even one of these diagrams.(48)   The creation of diagrams is linked to the excerpts and possibly to the creation of the excerpts.   One of the excerpts refers directly to an "appended figure" illustrating how a planet travels more rapidly through the sky when on the arc of its apsidal circle closest to the earth.(49)   It seems likely that the apsidal excerpt was the only one accompanied by a diagram at first, and that the success of the diagram led to the association of diagrams with the others as well. As we shall see very soon, such an assumption seems to require no further discussion in the cases of the order and intervals of the planets (excerpts 1 and 2).   However, the assumption invites further discussion for the other two excerpts.

There is clearly a pre-existing set of diagrams and some reason for confusion as early as the year 818 in CLM 210, where, in the three-book computus, the diagram for planetary intervals is already present.(50)   Unfortunately, this diagram follows immediately on the excerpt for the apsides and there is no apsidal diagram at all in the codex. In the very similar and contemporary manuscript, Vienna 387, probably copied from the same original as CLM 210, we find the same situation in the same place, but here the diagram for planetary intervals is incomplete.(51) What happened to the apsidal diagram, which was supposed to be there? It would seem that the original computus of 809 had already made the error, since both copies repeat it, which means that both the apsidal diagram and the intervals diagram - and most probably the other two as well - existed at the very beginning of the ninth century. The question then is, "how much earlier?" Let me suggest an answer of about 225 years, but only for the apsidal diagram. In a ninth-century **Aratea** manuscript at Leiden there appears a planetary configuration which can be dated to the year 579 and which is constructed on a pre-existing diagram for Plinian planetary apsides.(52) There is no good reason to date the excerpts so early, and we must conclude that the diagram predates the excerpt, which was constructed with both Pliny´s text and the Plinian apsidal diagram ready at hand.(53) The Plinian astronomical diagrams cannot be set in close parallel with the **rotae** of Isidore´s **Liber Rotarum,** for the Isidorean designs are most often clever mnemonics, while the Plinian figures are recognisably non-verbal representations of Pliny´s data. The conceptual bases of the two types are very far apart. With the apsidal diagram traced to late antiquity and the other diagrams predating the computus of 809, we are still left with no answer to the date of the excerpts themselves. In the absence of earlier exemplars we should conclude that they are Carolingian, **circa** 800, and that the accompanying diagrams appear at the same time, with the exception of the apsidal figure.

A similar question of dating emerges with regard to the latitude diagram. The early form of this design is a pattern of overlapping circles which appear to be a modified version of a sterographic projection. Who had such knowledge in Carolingian France or Germany? John North has commented on the knowledge of stereographic projection required to construct the diagram in CLM 210, f. 113v,(54) although it is not clear that this figure was invented when the text was written. The illustration may be a copy of an earlier one. With little foundation for speculation towards an answer, there seems, nevertheless, reason to avoid certitude in dating the earliest Plinian latitude diagram. We can only say that this type exists early in the ninth century.

Evolution of the diagrams

If the origins of these intriguing diagrams are shrouded, their basic transformation from Carolingian to post- Carolingian types is not. The change appears at about the middle of the ninth century and is neatly demonstrated by the diagrams in the manuscripts. Madrid codex 33-7, written sometime after 820, probably about 840, in Lotharingia (most likely Metz), shows the earlier stages.(55) Bern codex 347, written in the second half of the ninth century in France, perhaps Auxerre, shows the later stages of the diagrams.(56) The illustration for planetary order takes the simple, anticipated form of seven circles concentric to the earth, with each circle labelled in the Madrid manuscript (Fig.1), while the Bern version eliminates the orbits and simply places seven small circles, all equal except for the sun, in a horizontal sequence with labels (Fig.2). The same sort of thing happens to the diagram accompanying the excerpt for planetary intervals, where a set of concentric circles, proportionately spaced for their labelled harmonic intervals (Fig.3), becomes a vertical list of intervals with no circles at all (Fig.4).

The design for planetary apsides is the only one whose form seems to ramain fairly stable across the two manuscripts. This pattern probably underwent a much longer evolution, from a set of initially non-intersecting eccentric circles, gradually moving to intersections as in the Leiden configuration and others, for example Paris BN 5239, f. 125v (Fig.5), until all the circles intersected as in the Madrid manuscript (Fig.6) and the Bern manuscript (Fig.7). The only change of evident significance is the elimination of the stationary and retrograde arcs, a change better considered a variant rather than a new form. Completely intersecting circles is the final form, although one major variant on this pattern appears in the tenth century and will be mentioned later.

The diagram for planetary latitudes initially took a circular form (Fig.8), with the zodiac as thirteen concentric circles and each planetary path as an eccentric circle equal in size to the middle circle in the zodiac, viz. the sun´s path. The eccentricity of each planetary circle was set by the number of degrees that planet departed from the centre of the zodiac. Both the Madrid and the Bern manuscripts show this form (Fig.9), but the later codex also produces another design for latitudes (Fig.10), which may well have been invented at the time this manuscript was copied. The new form is a rectangular grid, 12 spaces by 13, with planetary paths weaving across from left to right with vertical intervals equal to the Plinian latitudes. This grid is an interesting translation of the circular latitude diagram, as the later design no longer has built into it a uniform longitudinal scale. One might think on first viewing that the horizontal scale of 30 spaces is a scale, but that number is the number of degrees in one zodiacal sign, and any attempt to reconcile planetary periods with the horizontal element of the grid shows quickly that no such scale or metric was ever in-tended. Instead, the number of completed longi-tudinal ´waves´ for each planet is (usually) inversely proportional to the amount of latitudinal travel, with the moon defining the basic unit.(57)

Overall, the diagrammatical history in these figures is more pedagogical than scientific. In every case - certainly in the latitude diagrams - there is no new breakthrough in conceptualisation of the astronomical data. There is only an increased simplification of presentation. With the changes in illustration of planetary order, intervals, and latitudes, the newer designs convey their information more quickly and eliminate other, secondary information that might be implied. This shows two things about the diagrams accompanying many, though by no means all, of the Plinian astronomical excerpts in the ninth to eleventh centuries. First, they are consciously attuned to teaching; these developments make sense only in the context of basic instruction. Second, they show an awareness of the value of conceptual abstraction in presenting new information. Each pattern shift, including the shift in apsidal diagrams from non-intersecting to intersecting circles, casts aside as much of the ancillary information as possible in the diagram, showing only one characteristic of the planets in each illustration and ignoring the relationship of the figure to the complex reality of planetary motion.

As a coda to this story of the diagrams we should attach an account of the evolution in the later tenth century in southern Germany of a new form of

apsidal diagram with re-ordered zodiac (Fig.11).
The pattern seems not to have spread much beyond its
origin.(58)   For the same pedagogical reason, the
zodiac of the diagram has only the six signs in
which the apsides are located.   These signs are
shifted round to achieve a better visual balance,
with distribution in every alternate thirty degree
segment being perhaps the most common result.(59)
However, this new distribution of signs (and their
apsides) is always intended to preserve the correct
sequence of the six with respect to each other.
The sole known violator of this precept was a
Spanish scribe, apparently inadequately informed of
the rules for constructing such a design.(60)
    By the eleventh century the Plinian astronomical
diagrams seem to have become second nature for many
scribes and even beyond any relationship to nature
for others.   Each of the four types of diagrams
begins to take on attributes of a decorative device
on the page by the eleventh century, and by the
twelfth century these diagrams can rarely be called
instructive at all.(61) (Figs. 12,13,14,15,16)

Pliny and Martianus in the ninth century
Having discussed the Plinian astronomical diagrams
from the ninth century to the twelfth, we have to
return to the ninth century for another set of
diagrams, associated with Martianus Capella.   The
history of the Capellan idea of circumsolar
planetary motion is outside our concerns, with one
exception.   At times there are readings of this
idea which explicitly associate Pliny with it.
The connection with Pliny occurs together with a
difficulty in the text, which ninth-century gloss-
ators found to have multiple possibilities.   The
manuscripts of Martianus´ book VIII show two major
traditions in the description of Mercury´s and
Venus´ paths round the sun.   One tradition gives
apparently concentric circles around the sun, while
the other results in intersecting circles around the
sun.(62)
    According to one manuscript tradition, Martianus
wrote "When both planets (Mercury and Venus) have a
position above the sun, Mercury is closer to the
earth; when they are below the sun, Venus is closer,
inasmuch as it has both a ´chaster´ (**castior**) and a
more open orbit".(63)   Whatever his reason for
using the word **castior**, it seems certain that the
author intended two concentric and heliocentric
circles here.(64)   By the time of the commentary of

Remigius of Auxerre, composed about 880, it seems to
have been generally agreed that **castior** meant
"closer". This has the passage describing non-
intersecting and presumably concentric circles in
its first part, then suggesting that the planet
which is supposed to be the outer one is not so at
all times, that is, not when **castior**.(65) Faced
with this situation, a reader might respond with an
emendation, as did the glossator to a ninth century
Leiden manuscript, noting marginally

> If he wanted to set out the planetary order
> according to Plato, the word "earth" in this
> sentence can remain. But if we wish to assume
> the order of planets according to the Pyth-
> agoreans and Pliny, we shall never have
> comprehension unless "earth" has been dropped
> so that the sentence reads, "but when they are
> above the sun, Mercury is closer", and the
> phrase "to the sun" is understood.(66)

Furthermore, the same manuscript has the word **terris**
crossed out in the text and an **ei** entered directly
above, referring to the sun, along with other,
supporting glosses to the passage.(67)

Lest we wonder what the Plinian version here
looks like, another ninth-century Leiden manuscript,
already containing the emendation just described,(68) goes further and adds a diagram. In
fact, three diagrams are added, and these three
diagrams appear to be directly consequent to the
prescriptions of the marginal gloss in the earlier
manuscript.(Fig.16)(69) On the right side of the
triple-version diagram is the pattern said to be
Plato´s by the glossator but here attributed to
Bede; it is a set of concentric circles centred on
the sun. At the centre is the pattern supposedly
of the Pythagoreans, and it is labelled **Martianus.**
With regard to the planet Venus, Martianus says
elsewhere approvingly that Pythagoras and his
followers studied the planet thoroughly,(70) and so
we are meant to understand Martianus and Pythagoras
to propose the same pattern. Finally, there is
Pliny´s pattern on the left, labelled with his name.
The Plinian and Capellan versions are linked
together by the gloss, and we can see that they have
in common an intersection of the two orbits around
the sun, as required for the emended text.
However, the Capellan version has full, circular
paths around the sun, while the Plinian has a pendant
truncated oval for Mercury and a pendant truncated

circle for Venus.

How was the Plinian pattern reached? If one reads Pliny´s account of his notorious **apsides conversae**(71) in the light of the Capellan text, one will find that Venus and Mercury are in every sense opposite to the superior planets, that this pair has its circles completely below the sun, and that they never cross the sun´s path. Combining these prescriptions with the commonly known facts of bounded elongation for Venus and Mercury, with Venus extending forty six degrees and Mercury only twenty two according to Pliny, we are constrained in our amalgamation of Pliny with Martianus to come to the design produced in the triple version. Here we see the full force of the incorporation by Martianus of Pliny´s heliodynamic astronomy. A Plinio-Capellan system, as it were, has been recognised by the ninth century readers of the two authors. Incidentally, this recognition requires more of Pliny´s text than is found in the astronomical excerpts. The crucial description of converse apsides (II.72-4) is not there. Among possible sources, one still extant is the Pierpont Morgan manuscript from the early or middle ninth century.(72)

The connection of Pliny with Martianus in the ninth century can also be seen in their amalgamation in comments on other texts. For example in a later-ninth century (c.876) Melk manuscript from Auxerre of Bede´s **On Time Reckoning**(73) there appears a contemporary comment on chapter 8 saying that Mercury and Venus have bounded elongations from the sun because they circle the sun as centre. The commentator continues with a close paraphrase of the **Natural History** regarding the limited **apsides** of these two planets.(74) Here Pliny is made to serve the Capellan commitment to circumsolar Mercury and Venus.

Final assimilation

Copying of Plinian texts, especially the astronomical excerpts and their diagrams, continued unabated through the tenth century. By the end of the tenth century a considerable body of traditional descriptive astronomy was not only availale but was being actively abbreviated and excerpted. This included texts of Macrobius, Martianus, Bede, Pliny, Hyginus and the **Aretea**. Abbo of Fleury, who flourished at the end of the tenth century, (75) composed computistical texts and an interesting brief

defloratio of astronomy, best known by its incipit, **Studiosis Astrologiae.**(76)   Abbo made extensive use of the Plinian excerpts and recombined some of the material to suit his own emphases.   He put together data on planetary latitudes and longitudinal elongations from the sun in order to emphasise that Mercury and Venus with bounded elongations had greater altitudinal variations than Mars, Jupiter and Saturn, with their unlimited elongations. Re-emphasising an idea of Pliny, Abbo pointed to a balance between longitudinal and latitudinal variations.   **Absides** Abbo defined as the places where the planets were installed at the Creation, following Macrobius´ **Commentary** on the specific location of each planet.(77)   He then shifted immediately to the planetary colours, which signify the effects of the planets, and described Mars, Jupiter, Venus and Mercury in Plinian terms.(78) After further description of planetary elongations and a lesson on how to tell a fixed star from a planet - for those willing to wait six months - Abbo presented a fine modern version of the rectangular grid pattern (12 by 12) for planetary latitudes. An overturned column-list of planetary harmonic intervals, the second Plinian excerpt, appears directly below the grid.(79)   It seems that Abbo invented this combination of the Plinian latitude and intervals diagrams.   He also included in his version of the Plinian excerpts with supplements an explicit reference to the heliodynamic cause of stations and retrogradations, "which the philosophers hold to be caused by solar rays".(80)

By the end of the tenth century the Plinian astronomical material had been fully digested, reformulated and integrated into the Latin educational tradition.   The tendency beginning in the eleventh century to use Plinian astronomical diagrams as decorative devices testifies not only to the absorption of these materials but probably to their obsolescence as well.   At the end of the eleventh century a commentary on Macrobius(81) suggested a relevance of Plinian apsides to the problem of the appearances of the inferior planets. Macrobius had remarked that Mercury and Venus seem to be above and below the sun, although he considered the truth to be a fixed suprasolar order.(82)

The commentator equated Macrobius´ superior vertices and the planets´ orbits with Plinian superior apsides, or apogees.   He then translated the Macrobian inferior arcs into the parts of Mercury´s and Venus´ circles which occur prior to

intersection of the circles (**internexio circul-
orum**).(83) The meaning of this phrase is not
obvious, but the equation of apsides with vertices
might well lead us to Pliny and his reference to
intersections or apsides (**commissurae apsidum**) when
he set reasons for planetary occultations.(84)

A much stronger connection between Plinian
astronomy and the Macrobian commentator emerges
where he has conflated apsides and latitudes: "the
apsides on the sun do not much exceed the two lines
nearest the ecliptic. Mercury does not cross the
full width of the zodiac. Venus however goes
beyond the zodiac".(85) The conjunction of apsides
and latitudes in the mind of the writer may have
come from a misunderstanding of Abbo´s **Studiosis
Astrologiae,** which tried to lay out relationships
between apsidal and latitudinal motions. Others
besides this Macrobian commentator fell into similar
confusion, witness the appearance in one manuscript
of Abbo´s work of an apsidal diagram in the place
where a latitude diagram was required.(86)

A final example of the fate of Pliny in this
transitional period appears in a mid-twelfth century
English manuscript, in which extracts from Abbo,
Adelbold, Bede, Isidore, Martianus, Pliny and others
can be found.(87) **Florilegia** are not unusual and
not usually remarkable, but the happy fate of a
Plinian astronomy fully assimilated finds good
witness here and deserves at least a brief notice.
Oxford Digby 83, in combining so many materials
makes some interesting choices. Apsides are again
called the places where the planets were created,
but this compiler locates each planet´s apsis at
mid-sign in the Plinian astrological apsides
(exaltations) rather than apogees.(88) He uses
Abbo for the Plinian heliodynamic phrases(89) and
goes on with a wonderful amalgam of sources, listed
here with the additional note that word changes,
transitions, and various forms of excerpting make
this a nice example of the results of the
descriptive tradition in medieval Latin astronomy to
the twelfth century. In the course of a page and a
half, the author quotes or uses the following:

f.34v, 17-26   **Natural History** II.61 (complete)
  34v, 27   **Natural History** II.69 (Beaujeu
30, 11)
  35r,1-5   Martianus, **De Nuptiis,** VIII.
854-5 (Dick 449, 26-450,5)
  35r,5-8   "De absidibus earum"   (Ruck 39,

1-4)

    35r,9-15   unlocated   material   from Greco-Arabic source(s)

    35r,16-17   **Natural History** II.79 (Beaujeu 34,6-7) **or** Bede **DNR** 15 (Jones 207,9)

    35r, 18/19-21/22-25 "De absidibus earum" (Ruck 39,10-17)

    35r,25-26   **Natural History** II.65 (Beaujeu 28,19-22)

    35r,27-35v,5   "De absidibus earum" (Ruck 39, 19-40,7)

Displacement in the twelfth century

Oxford Digby 83 testifies to a continuity of Plinian astronomy up to the twelfth century. At this juncture the work of William of Conches bears witness to the displacement of Pliny´s heliodynamic astronomy. From the mid eleventh to the mid twelfth century Plato´s **Timaeus** with Calcidius´ **Commentary** was one of the most important sources for Latin European cosmology and astronomy.(90) Along with the translations by Constantine the African and Alfanus of Salerno,(91) Plato and Calcidius provided William of Conches (c.1128) with his element theory, (92) which proved more than adequate replacement for the incomplete skeleton of Pliny´s heliodynamic system. William Shelley, as one English cataloguer was wont to call him, very neatly set aside the Plinian concept. While retaining a Plinian description of the stations and retrogradations, he replaced the Plinian cause, an inexplicable force of solar rays, with the more reasonable effects of solar heat. Here is the way he put it.

> Certain persons say that there is some point in the circle of any given planet at which, when the planet arrives, the sun makes it stand and perhaps also regress. But they do not say how this happens. We maintain that they never stand, but appear to, because, whereas their fiery nature requires them always to be in motion, they seem nevertheless to stand as a result of rising and falling, that is, elevation and depression. All observers of the stars (**astrologi**) agree that any planet is sometimes farther removed than usual from the earth and is then elevated, and sometimes descends closer to the earth than usual, which we call depressed. When they are elevated or depressed, if this happens on a line perpend-

icular to us, for when elevated they are seen
under the same degree of a sign, they are
believed to stand still. Or sometimes they
appear to return backwards. The cause of this
elevation and depression is the sun. Since it
is the source (**fons**) of all heat, sometimes it
dessicates the higher and lower bodies more
than usual. Thus dried out, the bodies of the
planets, lighter than usual, rise up. If,
again, in nourishing itself the sun attracts
more moisture than usual, it renders the
planets wetter and heavier than usual, so they
descend more than usual. That they are
therefore said to stand is an astrological
statement, because it is an appearance.(93)

Elsewhere William employs a rectangular pattern (but
not a grid) for the zodiac, along which he locates
the sun as the central horizontal (not serpentine)
and the moon´s path as a zigzag that crosses the
full twelve degrees of the zodiac.(94) But this
hardly qualifies as an "influence" of Plinian
astronomy - perhaps better a disconnected footnote
to Plinian astronomy.

The Plinian and then the Plinio-Capellan astronomy contained a central idea accounting for the
cohesion of the planetary system. Solar radial
force was integrated as much as possible with
Plinian data on the apsides and latitudes of the
planets by writers like Abbo of Fleury to emphasise
the interconnected balancing of planetary motions.
The idea of symmetry and balance pervaded the ninth
and tenth century revisons to the Plinian diagrams
for apsides and latitudes, also. The revival of
Platonic and Galenic element theory(95) spelled the
doom of Plinian astronomy even before the Ptolemaic
mathematical theory and Aristotelian dynamics rose
to primacy in the later twelfth century. William
put it well enough when defining the astronomical,
as opposed to fabulous and astrological, manner of
presenting knowledge. Astronomy, he said, considers that which is real, whether visible or not;
Ptolemy and Julius Firmicus Maternus exemplify this
kind of knowledge. Astrology on the other hand
concerns the appearances in the heavens, whether
real or not. William´s examples of the **astrologi**
are Hipparchus and Martianus Capella.(96) The
Plinio-Capellan celestial dynamics has thus been put
aside, preserving only the descriptive matter in
this tradition. The flood of Arabic learning via
Abu Ma´shar(97) and mid-century writers like Hermann

213

of Carinthia did not cause, but followed, the displacement of Plinian astronomy in Latin Europe.(98)

Plinius redivivus: renaissance commentators
The twelfth century renaissance had its positive
effects on manuscripts of Pliny´s **Natural History** as
on those of many other classical Latin authors.
Excerptors of Pliny worked on the text during the
century and show use of the older and more complete
manuscript tradition.(99) Unabbreviated English
manuscripts of much or most of Pliny written in the
twelfth century likewise show a good connection
with better exemplars from the pre-Carolinian
period.(100) Finally, it was the twelfth century
which sorted out the disorder introduced to a number
of manuscripts of the ninth to eleventh centuries,
re-establishing the proper sequence of material in
books I-V.(101) With this in mind we need not
become too excited over Petrarch´s dissatisfaction
at the corrupt state of a Pliny codex he bought in
Mantua in 1350.(102) Petrarch presumably came into
possession of one of the less dependable manuscripts
among those existing at his time.

One index of currency is of course the number of
surviving manuscripts. and it has been common to
repeat Detlefsen´s number of about 200 known
manuscripts.(103) Charles Nauert´s superb survey
of the literature on Pliny notes that the number
could be higher.(104) The number is indeed higher
if we count the larger number of florilegia of the
fourteenth and fifteenth centuries. However,
limiting our count to excerpts (brief or long) prior
to 1200, continuous texts of longer sections of the
Natural History, and complete texts, we can still
arrive at a conservative total of 180 manuscripts.
Inclusion of the later excerpts would push the
number well above 200. This puts Pliny´s **Natural
History,** judged by numbers alone, in the same
category as Martianus Capella´s **Marriage of
Philology and Mercury** and Macrobius´ **Commentary on
the Dream of Scipio.**(105) More to the point,
distribution by centuries shows that, far more than
these two competitors, Pliny is a renaissance text,
for more than half the manuscripts located have been
dated to the fifteenth and sixteenth centuries.
Renaissance printings would tell the same story,
with Pliny first printed in 1469 at Venice and at
least seventeen incunable editions.

Why the renaissance fascination with Pliny?
Without even considering the last thirty-five books,
we can say that book II is a unique, even if flawed,

encyclopedia of Hellenistic cosmology, especially the tradition apparently indebted to Posidonius via Varro. This murky window into the past appealed much more to those renaissance students of the varieties of classical knowledge than it did to the medievals, whose apparent taste for extracts may have been dictated by a preference for the excerptor´s clarification over the complexity and confusion of the complete text. On the one hand, for renaissance students of astronomy, Pliny offered only a primitive and debased knowledge when compared with the Ptolemaic tradition. On the other hand, Pliny´s astronomy was attractive, much more so than Martianus´, for composers of popular astronomical handbooks. Joachim Sterck von Ringelbergh of Antwerp included a great deal of Pliny along with Aretean and other materials in his non-mathematical **Institutiones Astronomicae** (1528, 1535).(106) A sample of the general renaissance commentaries on Pliny by Ermolao Barbaro (1492), Sigmund Galen (1539), and Jacques Dalechamp (1587) shows no ascertainable interest in the planetary theory and only the briefest of emendations, though usually quite laudable.(107) There are, however, nine commentaries before or during the first half of the sixteenth century primarily or solely on book II of the **Natural History,** and we turn now to these.(108)

Before looking at any of the nine commentaries in detail, we can observe that only three were composed with a concern both for the technical astronomy and the peculiar Plinian explanations of astronomical phemonena. We shall postpone consideration of these three. Of the others, one, that of the Anonymous Vaticanus (c.1550), does not proceed far enough into book II to reach the planetary astronomy. (109) Another, that of Petrus Olivarius (1536) is brief and was written to supplement earlier humanistic commentaries, continuing the concern with philological issues.(110) Georgius Valla´s commentary (1500-1502)(111) shows a concern to demonstrate the similarities of Pliny to other classical authors, although Valla does remark that Pliny´s account of the bounded elongation of the planets Mercury and Venus is inferior to Ptolemy´s explanation. Also, an epicyclic diagram is introduced to explain the morning and evening stations of a planet at one point, rather than the force of the solar rays, which was Pliny´s way of accounting for the phenomenon.(112) The commentary of Franciscus de Villalobos (1524) covers the preface and book II of Pliny, offering no noteworthy comments on the planetary astronomy.(113)

From the earliest commentary, by Bartholomeus Platina (c.1480)(114), to the last of the nine, by Paulus Eberus (1545/1556),(115) there appears some shift in attitude which has been said to characterise renaissance commentaries on classical authors in general. Charles Nauert points to a statement by Jacques Dalechamp in the preface to his commentary of 1587: "My mind had a natural proclivity to give precedence to those things that contribute to the understanding of the material rather than to the beauty and eloquence of speech..."(116) This preference of Dalechamp is noteworthy, but we may wonder how consistent he was in acting on it, for his very sparse notes on the planetary astronomy include a number of readings from a manuscript codex known as the Chiffletianus, and these readings are not always to be recommended, for example the elongations of Mercury and Venus as 20 and 36 degrees respectively, which is much further from the truth than the 23 and 46 degrees given in the text,(117) or the Chiffletianus reading of **sub terra** for the printed text´s **subter,** a suggestion which seems inferior on grounds of syntax as well as good sense.(118) On the other hand, these and other variant readings recorded from the Chiffletianus along with other manuscripts are interesting to the philologist and historian of the text. Keeping our focus rather sharply adjusted to Pliny´s astronomy, we can see that quarrels with Pliny´s account were already present at the beginning of the century, although it is clear that the later commentators disagreed more often with Plinian astronomy than did the earlier ones. With regard to astronomy alone, it seems that the degree of concern with truth as opposed to verbal fidelity to Pliny depended directly on the ability (or interest) in astronomy on the part of the commentator. For example, a vexed passage in the **Natural History** II.72, on **apsides conversae** was avoided or ignored by those commentators without some knowledge of astronomy while it was discussed at length by three German commentators who were trained in mathematical astronomy.(119) To Mauert´s argument "that a significant shift in attitudes towards ancient scientific texts was under way...The true goal of the natural philosopher was coming to be more clearly conceived as the determination of the truth about nature not just the determination of what a long-dead author had really written",(120) we may perhaps withold assent, at least from any extreme form of the interpretation, until more systematic study of the commentaries on Pliny has taken place. The shift seems so irregular and nuanced as to require further consideration of the disciplinary backgrounds of the individual commentators and de-emphasis of the apparent monolithic category of "commentator".

Of three German commentaries written in the 1530s by mathematicians, the earliest is that of Collimitius, or Georg Tannstetter von Thanau, a physician to the emperor Maximilian I and professor of mathematics at Vienna.(121) Published in 1531, the commentary may have been composed in part as early as 1519.(122) Collimitius´s **Scholia** on book II run to 44 pages, to which four pages on book XVIII were added. An intriguing association appears in the publication with these materials of the doctrines on the planets from Martianus Capella.(123) It would seem that a Plinio-Capellan astronomy continued to have some recognition.

Collimitius´ approach to Pliny has many things in common with his successors´ commentaries. All three assume a Ptolemaic model and explain Pliny´s astronomy in such terms. Their enterprise is not primarily philological but scientific and presumably intended for students. While Collimitius recognises and summarises Pliny´s heliodynamic, connecting it with Vitruvius,(124) there is mention made of it subsequently only in order to show its inadequacy. At one point the commentary says that Pliny´s mention of the force of rectangular solar rays on the moon when it is full can only make sense mystically.(125) Various corrections are made to Pliny´s assertions and data, such as the notion that apsides are fixed as well as the actual places of the apsides in Pliny´s time.(126) Along with some errors,(127) appropriate emendations appear which are, surprisingly, not retained in the critical edition of Detlefsen, used in the modern Loeb translation.(128) Reflecting perhaps some debate in his own time Collimitius remarks that progress, regress and stations of the planets caused by solar rays were not interpreted by Pliny according to epicycles, because his era either did not know or condemned that reasoning.(129)

Jacob Zeigler,(130) a humanist astronomer and geographer from Bavaria, educated at Ingolstadt, was something of a peripatetic in Italy and Germany, spending his last years at Vienna and Passau. While there are a few pages on related matters of books VI,XVII,XVIII and XX, Ziegler´s **Commentarius**

of 1531 concerns itself essentially (244 pages) with
the astronomy of book II. His dedicatory preface
shows an interest in Pliny as a precious source for
prisca astronomia(131) even though he then proceeds
to transform the contents into Ptolemaic explanations.(132) Ziegler remarks on the **spiritus**
which Pliny has keeping the planets and stars in
their places, then replaces this with the epicycles
and eccentrics of Ptolemy who, he says, does not
speak in the same manner as Pliny.(133) Because
Pliny's cosmology has planets with elemental
properties, and the properties of one affect others,
the commentary points out that these supposed
changes in the heavens raise difficulties,
and the peripatetics have a better view in regarding
the heavens as constant and any changes caused by
the planets to be only on earth.(134) In the
course of a long section (70 pages) on the planets,
Ziegler uses epicycles throughout. He equates
Pliny's **absis** with the modern astronomer's
epicyclus,(135) corrects Pliny's apogees(136) as
well as reforming the latitudes,(137) and describes
stations and retrogradations in good modern
form.(138) His attempt to explain Plinian stations
and retrogradations in their own terms involves a
comparison of the sun with Mount Etna, spewing out
rocks like planets, with the **vapor solis** preventing
them from falling to the earth. Vitruvius is said
to present a similar idea, using the rays of the
sun like the arm of a ballista to hurl the planets
away from the earth.(139) For all his ingenuity
however, Ziegler finds at least one passage, Pliny's
account of the converse apsides of the inferior
planets, to be quite inexplicable in its own terms.
The bounded elongation from the sun of Mercury and
Venus is without reason in the absence of epicycles,
and Ziegler gives up the effort of a non-epicyclic
account to return to the standard explanation.(140)

The last of the published renaissance commentaries on book II is that of Jacob Milich,(141) which
appeared first in 1535 and then in three more versions before his death in 1559. Born and educated
at Freiburg im Breisgau, he studied medicine and
mathematics at Vienna and taught at Wittenberg.
Milich supports astrology against its detractors and
praises Pliny's book II as the most suitable text
for "adolescents" beginning their studies in the
elements of astronomy and meteorology.(142) He
praises also Ziegler's commentary and obviously sees
Ptolemaic astronomy, with its modern improvements,
as the proper direction for more advanced students

of astronomy. He leads his reader in this direction by using epicyclic analysis wherever it will
make more sense of the text. Unlike the works of
Collimitius and Ziegler on Pliny, Milich devotes his
attention to the whole of book II and not to the
astronomy alone. Much of the text is closer to
paraphrase than to commentary.(143)

Milich´s **Commentaries** are essentially two, the
initial version of 1533 and then a serious refine-
ment presented in the almost identical versions of
1538 and 1543.(144) The revised version is not
much changed in the 1553 edition.(145) Initially
Milich built up the value of Pliny as a school text,
for example in the statement "Many volumes of
Aristotle are included briefly in this single small
book (by Pliny)".(146) Recognising that Pliny
never used epicycles for stations and retrogradations of the planets,(147) Milich can be less
than severe with his subject when improvements are
needed. Where Pliny gives three reasons for the
altitudo of a planet, the commentary softens the
difficulties by labelling each of these reasons with
a characteristic: the astronomical, the astrological
and the poetical, which is to say, common
opinion.(148) However, regarding Pliny´s account
of the force of the solar rays Milich says quite
directly,

> This whole passage has no notable usefulness,
> since Pliny´s reasons concerning the effects of
> the solar rays in drawing planets back and
> forth are quite wrong. The sun indeed acts
> on our inferior realms and the imperfections of
> mixed bodies (**corpora mixta**), but whether it
> acts in the same way on pure celestial bodies
> cannot be shown by any solid reason.(149)

By 1538 Milich has become rather less patient with
Pliny´s inadequacies. He discards the labels for
the three reasons for planetary **altitudo** and simply
says that Pliny makes no sense here without
epicycles.(150) As for the whole business of
heliodynamics, Milich retains his initial opinion,
beginning on a somewhat sharper note, "This reasoning of Pliny´s contains many absurd things".(151)

Given the rather clear statements of Ziegler,
Collimitius and Milich on particulars of Pliny´s
planetary astronomy, it would seem necessary to
propose a revision of the thesis that the use of
Pliny at Wittenberg by Philip Melancthon and his
successors as a text in natural philosophy was in

conscious opposition to the Catholic scholastic Aristotle.(152)   Nauert sees this use of Pliny as "part of the early Protestant attempt to reform the curriculum and to eliminate the authority of Aristotle".(153)   What our three German commentaries really show is the validity of only a part of the first half of Nauert´s description - "part of the early...(German) attempt to reform the curriculum".   The virtues of Pliny´s astronomy and cosmology were simplicity and variety of concrete examples as opposed to the more involved and abstract Aristotle.   Ziegler, we may recall, did not hesitate to commend the peripatetic cosmology by name when discussing the possibility of physical change in the heavens.   His rather picturesque analogies show his appreciation of Pliny´s concrete imagery, not of Pliny´s theoretical consistency, and Ziegler regularly replaced Plinian apsides with Ptolemaic epicycles, wherever the latter were clearer and more economical means of explanation. Likewise Milich pointed directly to Pliny´s superiority as a beginning text for "adolescents". Such awareness of the very limited superiority of Pliny over the theoretically denser Aristotle was indeed a basis for reform, but there seems no need to make it necessarily protestant, for there is neither explicit hostility towards Aristotle, nor uncritical acceptance of Pliny in these German commentaries.   As for "authority", Ziegler, Collimitius and Milich show that Pliny had no more authority than Aristotle and less than Ptolemy over and over again.   Pliny was simply considered more accessible than Aristotle for young students.

Clearly the astronomy of Pliny´s **Natural History** had its highest respect in the early middle ages, when it was treated as a valid science.   The heliodynamic theme gave it a sense of coherence which was appreciated at the time.   From the twelfth century onward this theme was considered invalid, but for various reasons the Plinian astronomy could still be found in twelfth-century primers on cosmology and in both the eleventh and the first half of the sixteenth centuries as an element of lower education.   Perhaps equally noteworthy, though it may be accidental, is the fact that the locus of interest in Plinian astronomy was essentially Northern, both earlier and later.   As a source for knowledge of astronomy, either scientific or historical, Pliny seems not to have been much consulted in his own land.

NOTES

*The research on which this study is based began during a year (1979-80) as NEH Fellow and member of the School of Historical Studies, Institute for Advanced Study (Princeton). Final stages of the work were supported by a grant from the National Science Foundation (SES 82-17726).
1.   I use the Budé edition as the preferred critical edition, esp. for book II.   See **Histoire naturelle, livre II**, ed. and trans. Jean Beaujeu Paris, Les Belles Lettres, 1950, pp.8-38 for sections 1-88
2.   Ibid., pp.xviii-xx
3.   Because a detailed investigation of Pliny´s stoic ideas is not part of my purpose here, I do not pretend to certitude on the point, but it seeems to me that the stoic doctrine about **tonos** fits rather well with the celestial forces mentioned by Pliny.   Regarding stoic **tonos** see Lutz Bloos, **Probleme der stoischen Physik**, Hamburg, Buske, 1973, pp 65-73;   David Hahm, **Origins of Stoic Cosmology**, Columbus, Ohio State U P , 1977, pp 169-173
4.   As Jean Beaujeu notes (**Histoire Naturelle**, 158, n.1) Pliny´s idea of active solar rays appears in Lucan, Censorinus, Vitruvius and seems to be a late Babylonian notion transmitted by Posidonius to Varro.
5.   Giovanni V Schiaparelli, **Scritti sulla storia della astronomia antica**, ed. Luigi Gabba, Bologna, Zanichelli, 1927, vol. III, p. 288, note.
6.   Martianus Capella, **De Nuptiis Philologiae et Mercurii**, ed. James Willis,, Leipzig, Teubner, 1983, pp. 302-337 for book VIII
7.   The most useful discussion of Martianus´ sources is the chapter in William Harris Stahl et al., **Martianus Capella and the seven Liberal Arts**, New York, Columbia U P, 1971, vol. 1, pp. 171-201, supplemented by the footnotes on sources in vol. II, 1977, pp. 314-344, where Stahl translates book VIII.
8.   In sections 884 (on Mars: ed. Willis 335) and 887 (all superior planets: ed. Willis 336-7).
9.   At 857 (ed. Willis 324) Martianus says that the centre of their circles is the sun; at 879 (d. Willis 333) he refers to their circles as epicycles.
10.   Theon of Smyrna, **Expositio rerum Mathematicarum ad legendum Platonem utilium**, ed. Eduard Hiller, Leipzig, Teubner, 1878, p. 186 (ch.33)

11. Martianus is most conservatively read in section 887 (ed. Willis, pp. 336-7) as referring to superior planets alone in explaining anomalies by solar force, but there is ambiguity in the phrasing, where "all the above" may refer to the inferior as well as superior planets.

12, Pliny, **Natural History,** II.64 (ed Beaujeu, 28)

13. Martianus, VIII.884 (ed.Willis 335.15-15), 885 (336.3) and 886 (336,8-9)

14. Pliny, **Natural History** II.65 (ed. Beaujeu, 28); Martianus, VIII.884 (335.18-19) gives Capricorn 29 for Mars; VIII.885 (336.4) for Jupiter; VIII.886 (336.9-19) for Saturn.

15. Martianus, VIII, 882-883 (334-335); Pliny,II.36-38 (Beaujeu, 18-19)

16. The basic studies of the manuscripts are those of Claudio Leonardi, especially his "I Codici di Marziano Capella", **Aevum,** vol.33, 1959, pp. 443-489; vol.34, 1960, pp. 1-99, 411-524. A summary view appears in my "Martianus Capella", in **Dictionary of the Middle Ages,** ed, Joseph Strayer et al., New York, Scribners, in press. The high point of interest in book VIII alone was in the 12th century. Commentaries on Martianus, including commentaries on book VIII alone, are located by Cora E Lutz, "Martianus Capella", **Catalogus Commentariorem et Translationum** vol.2, 1971, pp. 367-381; plus "Martianus Capella, addenda et corrigenda", ibid., vol.3,1976, pp. 449-452, by Lutz and John Contreni; two mss. not reported by Lutz are given by Nuchelmans in **Latomus,** 16, 1957, pp. 95-6. The continuing reputation of book VIII appears in Lambert of Auxerre, **Logica** (Summa Lamberti), ed. Franco P Alessio, Florence, La nuova Italia, 1971, p. 4, where **Astrologia Marciani,** is labelled as an adequate introduction of astronomy for those who want basic knowledge of the liberal arts. Lambert taught at the new Dominican convent of Auxerre in the mid 13th century.

17. Lucan, **Belli Civilis libri decem,** ed. A E Housman, Cambridge, Harvard U P, 1927, pp. 309-310 (X.201-203). Isidore of Seville, **Traité de la nature,** ed. Jacques Fontaine, Paris, CNRS, 1960, ch. 22, pp. 255/257.

18. Ibid., chs. 9-28. Use of Pliny is identified by Fontain at X.1.1; XXIII.2.12-13; XXIII.1.9; and XXVI.6.43-47.

19. Ibid., XXXV.1.1-5; XXXVII.1-5; XXXVIII.2,

17-18 and 4.32.
    20. Artur von Fragstein, **Isidor von Sevilla und die sogenannten Germanicusscholien,** Ohlau i. Schl.: H Eschenhagen, 1931, pp.37-8,82
    21. Ernst Maass, ed., **Commentariorum in Aratum Reliquiae,** Berlin, Weidmann, 1898, pp. 272-4.
    22. Paul Meyvaert, "Bede the scholar", in **Famulus Christi,** ed. G Bonner, London, SPCK, 1976, pp. 40-69, esp. 58-60. In addition, see Charles W Jones, "Bede´s place in medieval schools", ibid., pp. 261-285
    23. Bede, **Opera, I: Opera Didascalia,** ed. Charles W Jones, CCSL 123A, Turnhout, Brepols, 1975, pp. 173-234. I derive this division of major sources from Jones´ indications besides each chapter title of the major source or sources for that chapter.
    24. Ibid., p. 229, note; p.233, note. Jones also notes two places (Bede chs. 25 and 43) with similar material to that in the **Natural History** V and two places (Bede chs. 18 and 36) with material similar to the **Natural History** XVIII, but these four locations all have alternate, intermediate sources.
    25. Bede, ed. Jones, 202, 8-14; Pliny II. 105-106, ed. Beaujeu, 46-47
    26. Bede, chap. 12, ed. Jones, 204
    27. See above, n. 17, for the sources in Isidore and Lucan.
    28. Bede, chap. 13, ed. Jones, 204-205; **Natural History** II.32-39, 41,44,78,59
    29. Jones, 207; **Natural History** II. 63-4, 68
    30. In chapter 14 Bede uses sections 63-4 and 68 of Pliny; in chapter 16 Bede uses sections 66-7. Section 65 of Pliny contains the astrologocal **apsides,** which are not used by Bede.
    31. Bede, chap. 15, ed. Jones, 207; **Natural History** II.79. For planetary colour doctrines in astrology see Franz Boll (and Carl Bezold), "Antike Beobachtungen farbiger Sterne", **Abhandlungen der Bayerischen Akademie der Wissenschaft,** philos.-philol. u. hist Kl., Bd, 30, Heft 1, Munich, 1916, esp. pp. 19-28
    32. Bede, chap. 16, ed. Jones, 207-8; **Natural History,** II.66-7
    33. See, for example, the introduction by Jacques Fontaine to Isidore, **Traité de la Nature,** 73 - 80 specifically on the early Insular tradition of the **Liber Rotarum,** and the map following p. 83 on the diffusion of the

treatise.

34. Bede, **Opera de Temporibus**, ed. Charles W Jones (Cambridge, Mass.: Medieval Academy of America, 1943), p. 359. Bede, **Opera**, CCSL 123A, pp. 222-3 shows in the source notes the Isidorean bases for this Plinian material. See Karl Welzhofer, "Beda´s Citate aus der naturalis historia des Plinius", **Abhandlungen aus dem Gebiet der klassischen Altertums-Wissenschaft, Wilhelm von Christ... dargebracht,** Munich, Beck, 1891, pp. 25-41, e.g. 28 both pro and contra, but not in support of Bede´s access to any book of the **Natural History** beyond book VI. Regarding book XVIII, Welzhofer´s hypothesis (p. 37) seems **ad hoc** and unconvincing.

35. These two astronomico-computistical collections have been studied and greatly clarified with regard to their sources and compilation by Vernon H King, "An investigation of some astronomical excerpts from Pliny´s **Natural History** found in manuscripts of the earlier middle ages", B Litt thesis, Oxford, 1969, primarily pp. 1-79.

36. E A Lowe, "Codices rescripti: a list of the oldest palimpsests with stray observations on their origin", **Mélanges Eugène Tisserant,** 5 (Studi e Testi, 235; Vatican City, 1964), pp. 67-113, esp. nos. 3, 126, 157. A concise summary of the tradition of the **vetustiores** mss. appears in Karl Buchner, "Uberlieferungsgeschichte der lateinischen Literatur des Altertums", **Geschichte der Textuberlieferung der antiken und mittelalterlichen Literatur,** Zurich, Altlantis, 1961, vol.1, p. 407: 2 very incomplete texts, 5 fragments (3 palimpsests).

37. Above, n.24. Cf. Welzhofer, "Beda´s Citate", p. 40, for the most liberal number of direct quotations from Pliny that can now be admitted.

38. There have been many attempts to establish that an early ms. of the **Natural History** existed in England by the time of Bede and was the basis for Alcuin s knowledge of Pliny. Among these attempts are Welzhofer (above, n. 34); Karl Ruck, "Das Excerpt der Naturalis Historia des Plinius von Robert von Cricklade", **Sitzungsberichte der königlichen bayerischen Akademie der Wissenschaft,** plilos.-philol. u. hist. Kl., Jhrg, 1902, pp. 196-285; J Grafton Milne, "The text of Pliny´s **Natural History** preserved in English MSS", **Classical Review,** 7, 1893, pp. 451-452; Donald J Campbell, "Two MSS of the elder Pliny", **American Journal of Philology,** 57, 1936, pp. 113-123, of which Campbell´s mas. Phillipps 8297 (ms. C) is now New

York Pierpont Morgan ms. M871. More recently J Desanges, "Le manuscrit (Ch) et la classe des recentiores perturbés de l´Histoire Naturelle de Pline l´Ancien", **Latomus**, 25, 1966, pp.508-525, has reminded us that we have no example of the earlier (**vetustior**) mss. for **Natural History** I-VI except Leiden Voss, F.4, and eighth century Anglo-Saxon ms. which begins at II.196, beyond the astronomical sections; the Cheltenham ms. (Pierpont Morgan 871), written probably at Lorsch (s.IX), is the oldest of the **recentiores**, and it shows only a shuffling of four large sections, not a loss of material. In other words, the search for Bede´s text of Pliny continues but shows less and less hope of success.

39. V H King, above, n. 35, reviews these matters and gives a variety of substantiations from earlier modern sholars.

40. These excerpts were first studied with care and in detail by Karl Rück, **Auszüge aus der Naturgeschichte des C Plinius Secundus in einem astronomisch- komputistischen Sammelwerke des achten Jahrhunderts,** Programm des Königlichen Ludwigs-Gymnasiums für das Studienjahr 1887/1888, Munich, Straub, 1888, who called them the York excerpts and argued for an origin in the vicinity of Bede, an argument carried on by Welzhofer and others.

41. There are four drawn respectively from **Natural History** II. 12, 32, 34-36; 38-44; II. 83, 84; II. 59-61, 69, 70, 63, 64; II. 62, 66-9, 71, 75-8, 80, 78, 79, 76, 77. The excerpts were published by Rück, **Auszüge**, pp.34-43.

42. Leiden Univ. Bibl. ms. Voss. lat. F.4 (II), ff. 4-33. See above, n. 38. Description in Karel Adriaan de Meyier, **Codices Vossiani Latini, I: Codices in folio**, Leiden, Univ. Press, 1973, pp. 7-8.

43. The four are: Leiden Voss, Q. 43 (books II - XXXII.11) and Voss. F.1 (books I - XXXVII), London BL Arundel 98 (books I - XVIII), and Paris BN 6796A (books 1 - XXI); Montpellier Bibl. fac. médecine 473 has discontinuus excerpts from many books, no astronomy; Wolfenbüttel 160.1 Extr. contains book II alone.

44. Rück, **Auszüge,,** pp. 39-9; from Pliny, **Natural History** II. 69-70

45. Rück, pp. 39-40; **Natural History** II.63-4

46. Rück, p. 41; **Natural History** II.71

47. Rück, pp. 40-43; from many locations in Pliny, q.v.

48. It is also notable that the excerpts are never glossed. Likewise the longer mss. of Pliny

are rarely glossed and never extensively.

49. Rück, p. 40 (**De Apsidibus Earum**).

50. CLM 210, f. 123r.

51. Vienna NB 387, f. 123r.

52. See my article "Origins and contents of the Leiden planetary configuration (MS Voss. Q. 79, fol. 93v), an artistic astronomical schema of the early middle ages",**Viator**, 14, 1983, pp. 1-40, esp. 2-4 (dating), 10 - 18 (the apsidal portion of the configuration).

53. The late ancient creation of such **rotae** is a well known phenomenon in general, esp. with regard to wind-**rotae**, which had various radial divisions, including one into twelve parts. See Georg Kaibel, "Antike Windrosen", **Hermes**, 20, 1885, pp. 579- 624; Karl Neilsen, "Remarques sur les noms grecs et latins des vents et des régions du ciel", **Classica et Mediaevalia**, 7, 1945, pp. 1 - 113, esp. 107 - 108; J F Masselink, **De Grieks-Romeinse Windroos**, Utrecht,Dekker & Van de Vegt, 1956, pp. 133 - 139 on Pliny; Robert Böker, "Windrosen", **Realencyclopädie der classischen Altertumwissenschaft**, ser. 2, vol. 8, Stuttgart, Druckenmuller, 1958, coll. 2325 - 2381, esp. 2371 - 2381.

54. John D North, "Monasticism and the first mechanical clocks", **The Study of Time**, vol. II, ed. J T Fraser and N Lawrence, New York, Springer, 1975, pp. 381 - 398, esp. 386 - 387. An apparently direct copy appears in Vat. Reg. 123, f. 205r (AD 1056); a modified or indirect copy in Bern 88, f. 10v (s.X)

55. On this astronomical computistical collection see Wilhelm R W Koehler, **Die karolingischen Miniaturen**, Berlin, Verein f. Kunstwiss., 1960, vol.III, pp. 119 - 127. The same diagrams appear in Paris BN n.a.l. 1615, ff. 181r, 159v, 160v, 161r, a ms. probably from Auxerre (s.IX (1)).

56. On this collection of excerpts from Macrobius, Pliny, and Nonius Marcellus, once part of a much larger codex, see Otto S Homburger, **Die illustrierten Handschriften der Burgerbibliothek Bern. Die vorkarolingischen und karolingischen Handschriften**, Bern, Burgerbibl., 1962, pp. 134-136.

57. Many studies of this grid pattern exist, and none seems to have sorted out the elements satisfactorily. Probably the best of these studies is Harriet P Lattin, "The eleventh century MS Munich 14436: its contribution to the history of coordinaters, of logic, of German studies in France", **Isis**, 28, 1947-8, pp. 205 - 225, esp. 215 - 221.

Some preliminary findings, focused specifically on the diagrammatic elements, appear in my article, "MSS. Madrid 9605, Munich 6364 and the evolution of the two Plinian astronomical diagrams in the tenth century", **Dynamis. Acta Hispanica...**, 3, 1983, pp. 265 - 280.

58. See ibid. and my earlier article, "Characteristics of the Plinian astronomical diagrams in a Bodleian palimpsest, ms. D´Orville 95, ff. 25-38", **Sudhoffs Archiv**, 27, 1983, pp. 2-12, esp. 6-8.

59. Examples are CLM 14436, f. 60v; Zurich Car. C. 122, f. 41v; Bern 265, f. 58v; as well as Vat. Palat. Lat. 1577, f. 81v.

60. The copyist for Madrid 9605, f. 12r. discussed in art. cit. above, n. 57.

61. Baltimore Walters ms. 10.73, from the end of the 12th century, shows some nice examples of the results of the decorative trend. See Harry Bober, "An illustrated medieval school-book of Bede´s ´De natura rerum´", **Journal of the Walters Art Gallery**, 19 -20, 1956-7, pp. 64 -97. Examples for the diagrams I have been considering are: Paris BN 8663, f. 24r; Madrid 9605, f. 12r.; Vat. Regin. 123, f.169v. (circular), Madrid 9605, f. 12v. (grid); see Figs. 12-15 for these.

62. I have described many of the elements of these two manuscript readings plus the appended diagrams in "The chaster path of Venus (**orbis Veneris castior**) in the astronomy of Martianus Capella", **Archives Internationales d´Histoire des Sciences**, 32, 1982, pp. 145 -158. A thorough examination of the various traditions will appear in my study of the idea of sun-centred planetary motion in Europe from late antiquity to the renaissance.

63. This reading, common to many manuscripts, was still preserved in the edition of Ulrich F Kopp, Frankfurt, 1836, pp. 668 - 669; ed. Willis, 324, 17 (VIII, 857) emends a crucial word, **castior** to **vastior**.

64. I have offered a speculative explanation for the Capellan term in art. cit. above, n. 62, pp. 146 -149.

65. Remigius of Auxerre, **Commentum in Martianum Capellam**, ed. Cora E Lutz, Leiden, Brill, 1965, vol. II, pp. 275, 25-28.

66. Leiden Univ. Bibl. ms. BPL 88, f. 162v.

67. Ibid., f. 162v, 9-14.

68. Leiden Iniv. Bibl. ms. BPL 36, ff. 110v, 30 - 111r, 4

69. Ibid., f. 129r.

70. VIII, 882; ed. Willis, 334
71. **Natural History**, II. 72-4; ed. Beaujeu, 31-2
72. For this manuscript, see above, n. 38.
73. Melk 412; discussed as cod. 370 by Charles W Jones, "A note on concepts of the inferior planets in the early middle ages", **Isis**, 24, 1935-6, pp. 397 - 398.
74. Melk 412, p. 35; Jones, "A note", p. 398. **Curvatura ergo absidum eas amplius ire non promittit quoniam non habent longitudinem ad solem. Ergo reciprocantur a longissimis distantiae suae finibus.** This part of the comment clearly derives from **Natural History** II.72.
75. The standard biography is by Patrice Cousin, **Abbon de Fleury-sur-Loire**, Paris, Lethielleux, 1954; the basic study of Abbo´s scientific work is by André Van de Vyver, "Les oeuvres inédites d´Abbon de Fleury", **Revue Bénédictine**, 47, 1935, pp. 125 - 169.
76. Manuscripts of this text are numerous. Lists can be found in Lynn Thorndike and Pearl Kibre, **A Catalogue of Incipits**, rev. ed., Cambridge, Mediaeval Academy of America, 1963, 400, 1530; **Osiris**, 1, 1936, pp. 667 - 677; **Revue Bénédictine**, 47, 1935, pp. 140 - 142. I have consulted Baltimore Walters 10.73, ff. 5v-6r, 4r-v; Bruxelles 2194 - 2195, ff. 52v - 56r; Cambridge Trinity R 15.32, ff. 3r-6v; Dijon 448 (269), f. 72r, 1443; Florence Laur. Pl. 51.14, ff. 72r - 76r; Köln Schnütgen Mus. Ludwig XII.5, f. 103r-v (now Los Angeles, Getty); Leiden BPL 225, ff.23r-27v; London BL Cott. Tib. E. IV, ff.141r - 143r; Cott. Vit.A. XII, ff. 8r-10v; Egerton 3088, ff.83r-84v; Harl. 2506, ff. 30v-32r; Royal 13, A XI ff. 113r - 115v; Oxford St John´s 17, ff. 37v-39r; Paris 12117, f. 183v
77. Macrobius, **Commentarii in Ciceronis Somnium Scipionis**, ed. Ludwig Jan, Quedlinburg, Bassius, 1848, pp. 123-4 (I, xxi, 24). Abbo omits the sun; see, e.g., Cambridge Trinity Coll. ms. R 15. 32, f. 4r, 13-15.
78. Ibid., f. 4r, 15-19: **Quorum differentias potes cognoscere exsitu vel qualitate vel quantitate. Mars quippe quantitate mediocris, rubeus est et terribilis, Juppiter quantitate magnus colore clarus, mercurius supradictis duobus major et rutilando clarior, venus que et vesper et lucifer dicitur maior reliquis planetis iiiior valde splendens cum candore.** Cf. **Natural History** II.79; Bede, **DNR**, 15. Cicero, **Somnium** IV.2, refers to

Mars as **terribilis**.
79. Cambridge Trinity R. 15.32, f. 4v. Of the mss. I have inspected (above, n.76), all have the grid or no diagram with the exception of Florence Laur. Pl.51.14, f. 73r, which erroneously gives an apsidal diagram.
80. Trinity R.15.32, f. 4r, 10. The same clause and phenomena appear in Melk ms. 412, pp. 26b - 27a, from Auxerre (s. IX), used by Heiric.
81. Köln Dombibl. ms.199, ff. 26v-38v
82. Macrobius, **Commentary**, I.xix,6 (ed. Jan 104)
83. Köln 199, f. 35ra, 7-11. Alison Peden, "Science and philosophy in Wales at the time of the Norman conquest...", **Cambridge Medieval Celtic Studies**, 2 winter 1981, pp. 21 -45, esp. p. 40, notes the use of **absides** to explain vertices, perhaps intersecting and perhaps not, in a late 11th century gloss to Macrobius in London BL Cott. Faustina C 1, f. 79v.
84. **Natural History** II.79
85. Köln 199, f. 35ra, 12-15
86. Above, n. 79
87. Oxford Bodl. Digby 83, ff. 32v - 42v. Concerning this ms. see Fritz Saxl and Harry Bober, **Verzeichnis astrologischer und mythologischer illustrierter Handschriften**, London, Warburg Institute, 1953, III, 1, pp 345-6.; L Thorndike, **History of Magic and Experimental Science**, New York, Columbia U P , 1923, vol. I, pp. 705 - 709; **Osiris**, 1, 1936, pp. 689-91.
88. Digby 83, f. 35r, 18-19, 24-27: Saturn in Libra, Sun in Aries, Mars in Capricorn, Jupiter in Cancer, Mercury in Virgo, Venus in Pisces, Moon in Taurus; see **Natural History** II.65.
89. Digby 83, f. 34v.
90. Margaret T Gibson, "The study of the Timaeus in the XIth and XIIth centuries", **Pensamiento. Revista de investigación y información filosófica**, 25, 1969, pp. 183 - 194, esp. 190.
91. For these translations and relevant literature see the excellent survey of Marie-Thérèse d´Alverney, "Translations and translators", **Renaissance and Renewal in the Twelfth Century**, ed. R L Benson, G Constable and C D Lanham, Cambridge, Mass., Harvard U.P., 1982, pp. 421 -462, esp. 422 - 426.
92. For an introduction to William´s use of Plato see Guillaume de Conches, **Glosae super Platonem**, ed. Édouard Jeauneau, Paris, Vrin, 1965, pp. 9 -31. On William´ element theory see Theodore

Plinian Astronomy in the Middle Ages and Renaissance

Silverstein, "Guillaume de Conches and the elements", **Mediaeval Studies,** 26, 1964, p. 364; also his "Guillaume de Conches and Nemesius of Emessa", **Harry Austryn Wolfson Jubilee Volume,** Jerusalem, Amer. Acad. Jewish Research, 1965, II, pp. 719 - 734
   93. Wilhelm von Conches, **Philosophia,** ed. Gregor Maurach, Pretoria, Univ. S.Africa, 1980, pp.53-4 (II, 33-5. Again in his **Dragmaticon,** published as **Dialogus de Substantiis Physicis,** ed. Guilielmo Grataroli, Strasbourg, J Rihelius, 1567, p. 104.
   94. Ibid., p.140. Among the many mss. of the **Dragmaticon,** Vat. Regin. lat. 72, f. 96v (s.XII), shows the faithfulness of the published to the early ms. version of this diagram.
   95. Richard McKeon, "Medicine and philosophy in the eleventh and twelfth centuries: the problem of elements", **The Thomist,** 24, 1961, pp. 211 - 256, esp. 253-5
   96. **Philosophia, II. 9-10** (ed. Maurach, p.44)
   97. Richard J Lemay, **Abu Ma´shar and Latin Aristotelianism in the Twelfth century. The Recovery of Aristotle´s Natural Philosophy through Arabic Astrology,** Beirut, Amer. Univ., 1962, esp. intro. and pt. 1
   98. Hermann of Carinthia, **De Essentiis,** ed. Charles Burnett, Leiden, Brill, 1982, pp. 1-43. Hermann´s cosmology is strongly influenced by the **Timaeus,** also.
   99. Among twelfth century sets of excerpts making use of the mss. **vetustiores** are Montpellier Ecole de Médecine ms. 473 (see D J Campbell, "A medieval excerptor of the elder Pliny", **Classical Quarterly** 26, 1932, pp. 116 - 119) and Robert of Cricklade´s compression of Pliny´s 37 books into 9 books (see Karl Rück, "Das Exzerpt... von Robert von Cricklade", n. 38 above, esp. pp. 216, 264).
   100. London BL Arundel 98 (bks. 1-18), Oxford New College 274 (bks.1-19); J Grafton Milne, "The text ... in English MSS.", n. 38 above.
   101. J Desanges, "Le manuscrit (Ch)...", n. 38 above, p. 520
   102. Petrarch´s emended and annotated copy is Paris BN lat. 6802. See Marjorie Chibnall, "Pliny´s **Natural History** and the middle Ages", in **Empire and Aftermath: Silver Latin II,** ed. T A Dorey, London, Routledge and Kegan Paul, 1975, pp. 57 - 78, esp. 74, 77.
   103. Detlef Detlefsen, **Die geographischen Bücher (II, 242 - VI Schluss) der Naturalis Historia des C Plinius Secundus,** Berlin, Weidmann, 1904, p.v.

230

104. Charles G Nauert, Jr., "Caius Plinius Secundus", **Catalogus Translationum et Commentariorum**, 4, 1980, pp. 297 - 422, see p. 304, n. 20.

105. I have made my own preliminary census of Macrobius and found over 200 mss. I have in progress a census of Pliny mss. and can guarantee at least the 180 mentioned. Claudio Leonardi, n. 16 above, has made an excellent survey of the Capellan mss.

106. Ringelbergh, Joachim Sterch van, **Institutiones Astronomicae ternis libris Contentae**, Basel, Curio/Koln, Quentell, 1528; Venice, Antonius de Nicolinis de Sabio, 1535, does at least refer to recent authors, e.g. f. 28r, where the **orbes planetarum** are mentioned in connection with Purbach.

107. Ermolao Barbaro, **Castigationes Plinianae**, Rome, Eucharius Argenteus Germanus, 1592, ff. a ii(r), d i(v) as examples. (Sigmund Gelen) Pliny, **Historia Mundi**, Basel, Froben, 1539, f. aaa i(v). (Jacques Dalechamp) Pliny, **Historia Mundi**, Lyon, Bartholomaeus Honoratus, 1587, pp. 9-11 for ch.s. 17-17. A very full list with bibliographical data for 46 genuine editions/commentaries can be found in Nauert´s survey, n. 104 above.

108. Nauert, "Plinius", is the essential source for these, to which further specifications are added below. He identifies a total of 12 commentaries up to that of Goclenius (1612) which are concerned solely or extensively wih book II.

109. The Anonymous Vaticanus, a commentary on **Natural History** II. 31-58, appears in Vat. lat. 3436. See Nauert, "Plinius", pp. 399 - 400.

110. Olivarius´ commentary covers **Natural History** II. 1-84. See Nauert "Plinius" , pp. 390-1.

111. The commentary attributed to Georgius Valla was published at Venice in 1502, and it deals with the preface and book II. See Nauert, "Plinius", pp. 350-1. As Nauert observes "the work was written down by the son from his father´s dictation of materials from his lectures on Pliny but was prepared for publication by the son after the father´s death in 1500" (p.350). The same commentary appears in manuscript form in Bologna Bibl. comunale dell´ Archiginnasio, ms. A 78, ff. 78, and it is atributed to the son, Johannes Petrus Valla, in the summary catalogue by C Lucchesi, in the **Inventari** of Mazzatinti and Sorbelli, vol. 30, 1924, p. 44. J P Valla was J P Cademustus, who wrote a life of his adoptive father, Georgius Valla, published notes on (Venice, 1516) and an edition of (Venice 1498, 1511) Plautus , and published the

Logica of Nicephorus Blemmydes (Venice, 1498). See P O Kristeller, **Iter Italicum,** London, Warburg Institute, 1967, vol. II, pp. 364, 499, and M E Cosenza, **Biographical and Bibliographical Dictionary of the Italian Humanists,** Boston, Hall, 1962, vol. IV, p. 3549. As Nauert observed, the commentary would seem to be the product of both men, father and son, and the Bologna ms., which is apparently the same as the 1502 publication, can reasonably be dated 1500 - 1502, though not with certainty.

112. Bologna Bibl. comunale dell´Archiginnasio ms. A 78, ff. 47r, 37r - 38v, respectively. Valla does not summarise Pliny´s doctrine of solar radial force with regard to stations and retrogradations of the planets (f. 44r-v). He also says (f. 42v) that the **mathematici** use epicycles to explain apsides. Among the comprehensive commentaries on the **Natural History,** that of Stephanus Aquaeus (Paris, Galliotus Pratensis, 1530) ff. 25v - 26r, also introduces epicyclic analysis for planetary motions. Valla´s sole exposition in any detail on Pliny´s radial solar force is in response to Pliny´s statement **Percussae in qua diximus parte et triangulo solis radio inhibentur rectum agere cursum** (**Natural History** II.69). Valla repeats this statement and continues (f. 44r-v)

> In hac eadem est Vitruvius opinione, inquens ergo potius ea ratio nobis constabit quod fervor quemadmodum omnes res evocat et ad se ducit ut etiam fructus ex terra surgentes in altitudinem per calorem videmus, non minus aquae vapores a fontibus ad nubes per arcus excitari; eadem ratione solis impetus vaehemens radiis trigoni forma porrectus in sequentes stellas ad se producit, ante currentes veluti re/frenando (f.44v) retinendoque progredi, sed ad se regredi in alterius trigoni signum esse...

113. Nauert, "Plinius", pp. 361-3
114. Platina commented on bks. II - V; his commentary appears in London BL ms. Harl. 3475, ff. 40. s.XV. See Nauert,"Plinius", pp. 335-7.
115. Paulus Eberus, scholar and teacher at Wittenberg, received a professorship there in 1544 and made Pliny´s work the subject of his first lectures. Nauert, "Plinius", pp. 400 - 402, gives valuable information, reporting one manuscript of Eber´s commentary on book II, Vat. Palat. lat. 1560, 1556. However, there exists an earlier dated

manuscript, Krakow Bibl. Iniw. Jagiellonski ms. lat. 3222 (2859), pp. 570 + 2, dated June 1545. This codex, which contains two printed texts dated 1544 as well as manuscript texts dated 1545, was owned (and perhaps written) by Isaac Schaller of Ernfridsdorfen. Pages 469-570 contain **In librum Plinii Annotationes, a mgro Paulo Ebero, Witebergae;** pp. 525-560 cover **Natural History II. 59-80.** Eber showed in these annotations no interest in philological or literary interpretation and did not discuss Pliny´s notion of solar radial force, contenting himself with transforming Pliny´s discussions into Ptolemaic explanations of the phenomena concerned.

116. Charles G Nauert, J., "Humanists, scientists and Pliny: changing approaches to a classical author", **American Historical Review,** 84, 1979, pp. 72 - 85, at p. 85.

117. Nauert, "Plinius", pp. 409-412, for Dalechamp´s commentary. For the reading specified here, see Pliny, **Historia Mundi libri XXXVII... novissime vel laboriosis observationibus conquista, & solerti judicio pensitata, Jacobi Dalecampii, Medici, Cadomensis,** Lyon, B Honoratus, 1587, p. 10 **(ad II.17 init.).**

118. Ibid., p. 10

119. The discussions of Ziegler, Tanstetter and Milich on Pliny´s "converse apsides" turn not only on ther truth of the matter but also on which of the two well-attested readings was the correct text of Pliny. The situation is reviewed in my article, "Kepler as Historian of Science: precursors of Copernican heliocentrism according to **De Revolutionibus** I, 10", **Proceedings of the American Philosophical Society,** 126, 1982, pp. 367 - 394, esp. 373 - 376.

120. Nauert, "Humanists, scientists and Pliny", p. 85

121. Bibliography and brief biography in Nauert, "Plinius", pp. 378 - 380.

122. Ibid., p. 378

123. Georgius Collimitius, **In eundem secundum Plinii Scholia** (published with Ziegler´s commentary), Basel, H Petrus, 1531, pp. 447 - 454, from **De Nuptiis** VIII, 875 - 887

124. Ibid., p. 418

125. Ibid., p. 426; ad **Natural History** II.80

126. **Scholia,** p. 421

127. Such as the interpretation of Pliny´s **absidum commissurae** as the farthest apogees (ibid., p. 425).

128. Such as the replacement of **sub terra** with **subter**, ther preference for which Collimitius explains in detail; see ibid., p. 435. This emendation was already suggested by Barbaro Ermalao in his 1492 edition (f.a ii(r): ch. 17). The text is II.72, and the Loeb translation (p.218) shows the difficulty that is caused.

129. **Scholia,** p.434 (**ad Natural History** II.65)

130. Nauert, "Plinius", pp. 375 - 378

131. Ibid., p. 375

132. Jacob Ziegler, **In C. Plinii de Naturali Historia librum secundum Commentarius,** Basel, H Petrus, 1531, p. 35, notes that Pliny made absolutely no mention of epicycles and deferents.

133. Ibid., p. 70

134. Ibid., p. 87

135. Ibid., pp. 152-3

136. Ibid., pp. 170-2

137. Ibid., pp. 179-86

138. Ibid., pp. 186-91

139. Ibid., p. 191

140. Ibid., pp. 193-8

141. Nauert, "Plinius", pp. 384-6. Unlike Nauert, I do not treat the modest reworking by Bartholomaeus Schonborn (publ. 1573) of Milich´s commentary as a new work. For my purposes, it does not contribute to the history of Plinian astronomy beyond witnessing to its continued use as a school text at Wittenberg. On Schonborn see ibid., p. 402.

142. Ibid., pp. 384-5, prints the latter part of the preface, in which Milich summarised the educational goals of his commentary.

143. Jacob Milich, **Commentarii in librum secundum Historiae Mundi C. Plinii,** Hagenau, Petr. Brubachius, 1535.

144. Under the title **Liber Secundus C. Plinii de Mundi Historia, cum Commentariis,** the second was published by P. Brubachius, Swabische Halle, 1538. Again by P Brubachius (Frankfurt, 1543), the next printing has the same pagination as in 1538 and the same wording in the spots checked; however, a few astronomical observations dated to the interim period have been added.

145. Still with the same title as 1538 and 1543, the Frankfurt, 1553 printing was also done by Petrus Brubachius. In many places, such as the account of **apsides conversae** of the lower planets, the wording is identical; see f. 69r-v (1543), p. 140 (1553). I have not yet found significant changes in the 1553 commentary.

146. 1535 ed., f. Ci(v)

147. 1535 ed., f. Mi(r)

148. 1535ed., f. Niii(r); **ad Natural History** II.64-5. Note that this humanistic device is similar to that of William of Conches (above, n. 96).

149. 1535 ed., f. Oi(v), 20-25; **ad Natural History** II.69-71

150. 1538/1543 eds., f. 61r-v; 1553 ed., pp. 124-5

151. 1538/1543 eds., f. 66v; 1553 ed., p. 135

152. Nauert, "Humanists, scientists, and Pliny", pp. 80-81

153. Ibid., p. 80. Nauert also says that this move at Wittenberg was "of small intrinsic importance", since Aristotle was reintroduced into the university curriculum even by Melancthon.

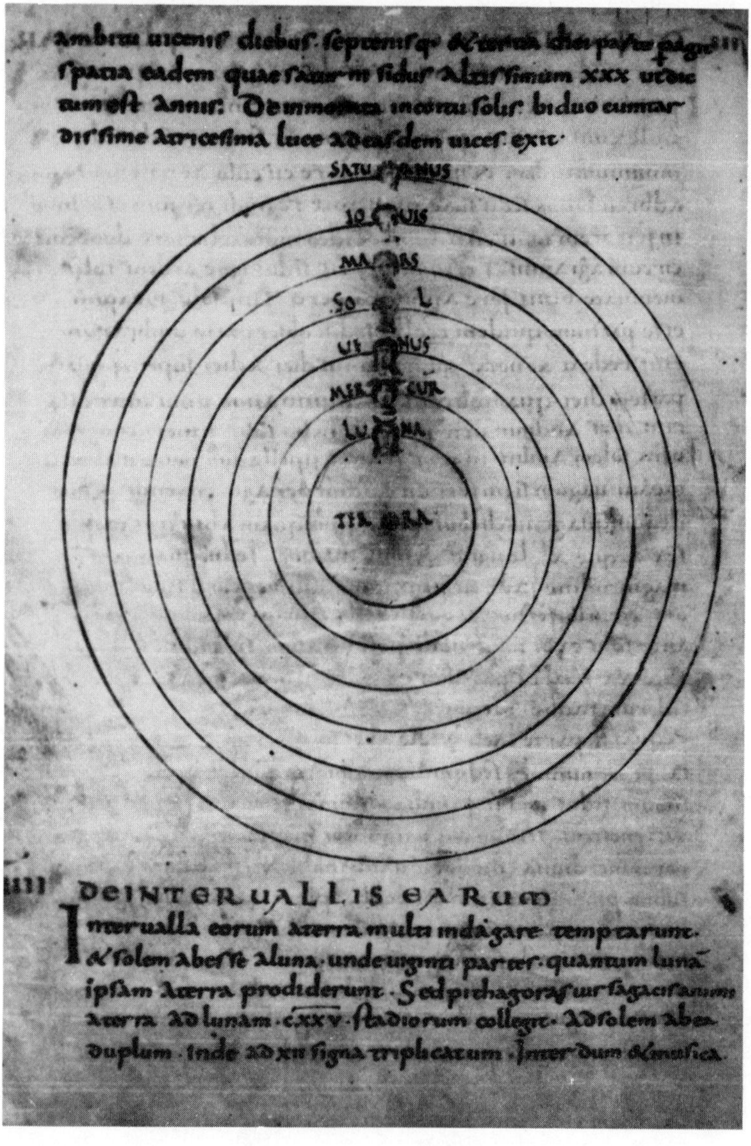

Figure Two

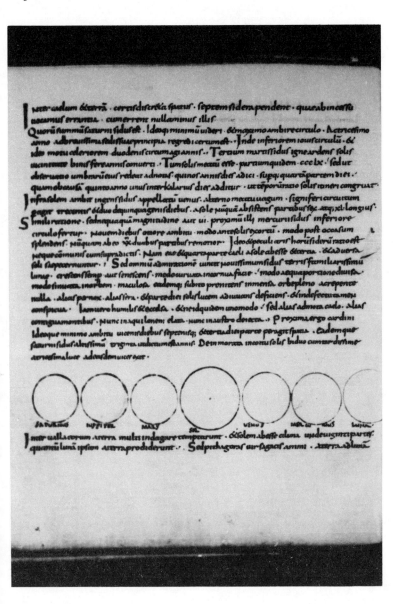

ratione appellat tonum quantum abstit a terra luna. Ab ea
ad mercurium dimidium spatii. hoc est semitonium. et ab eo
ad uenerem tantundem. A qua ad solem sescuplum. id est tria
semitonia. A sole ad martem tonum. id est quantum ad
lunam a terra. Ab eo ad iouem dimidium. et ab eo ad sa
turnum tantundem spatii. inde ad signiferum sescuplum.
ita septem tonis effici quam dia pason armoniam uocant.

SIGNIFER
SATURNUS
IOUIS
MARS
SOL
UENUS
MERCUR
LUNA
TERRA
TON
SEMIT
SEMIT
TRIA
SEMIT
TONUS
SEMIT
SEMIT
TRIA
SEMIT

Figure Four

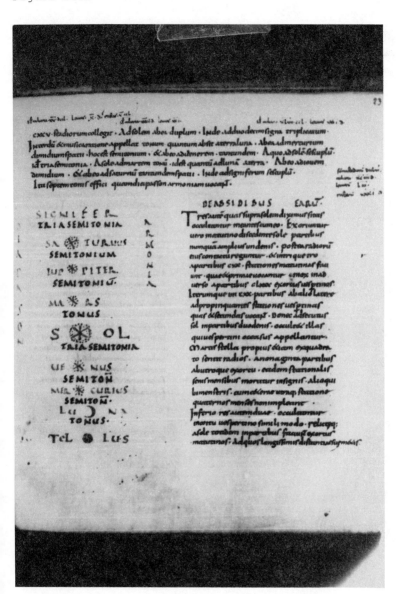

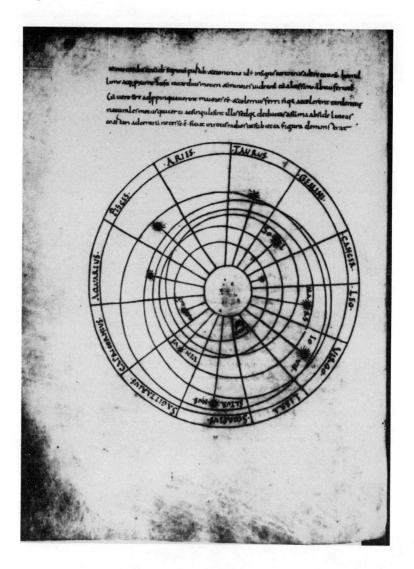

Figure Six

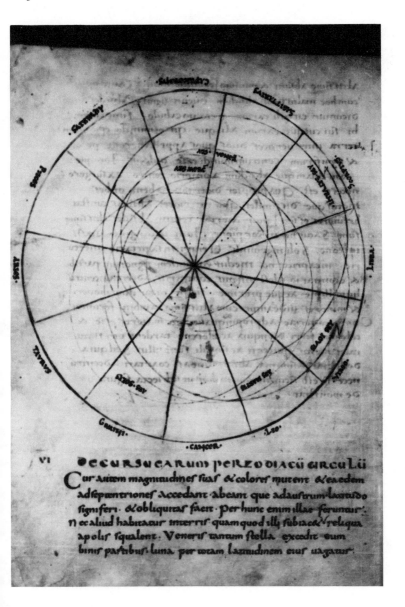

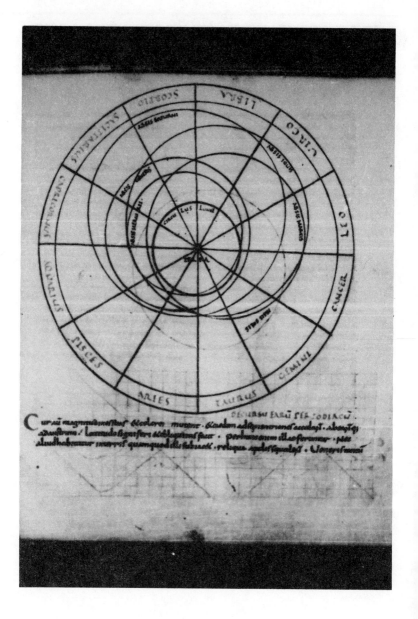

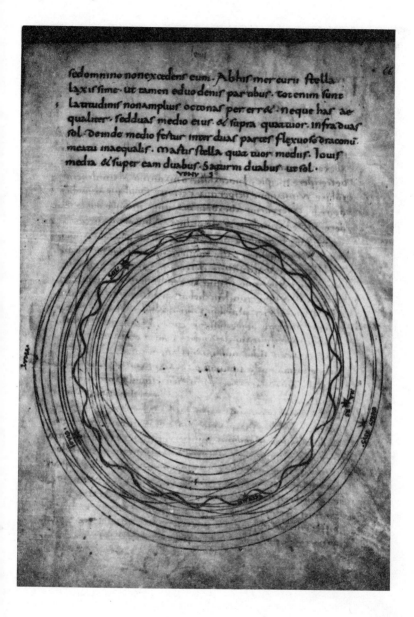

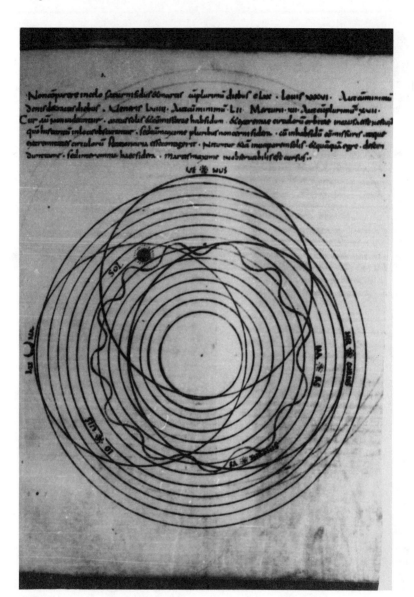

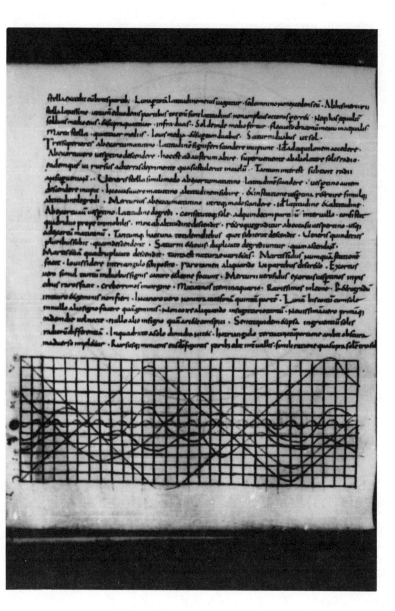

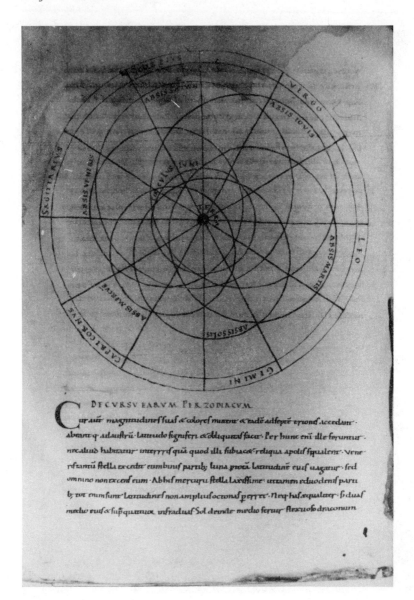

DE CVRSV EARVM PER ZODIACVM

Cur autem magnitudines suas & colores mutent & eade adseptre tipone accedant
abeant q adaustru latitudo signiferi & obliquitas facit. Per hunc em ille feruntur
nec aliud habitantur interrss qua quod illi subiacet reliqua apolis squalent. Vene
restantu stella excedit eumbinus partib; luna prota latitudine eius uagatur. sed
omnino non excens eum. Abhis mercuru stella laxissime uttamen eduodeni parti
b; tot emin sunt latitudines non amplius octonas perret. Neq; has aequalater. si duas
medio eius & sup quatuor. infraduas. Sol deinde medio seruit flexuoso draconum

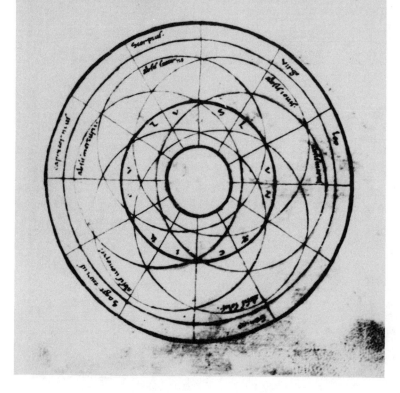

tibus tot enim sunt latitudinis · non amplius octonas per errat · Neq̄ · has equa
lis sed duus medio eius · et supra quattuor infra duus · Sol deinde medio fertur
medius partes · flexuoso draconis meatu inequalis · Martis stella quattuor
mediis · Iouis media et supeam duabus · Saturni duabus ut sol ·

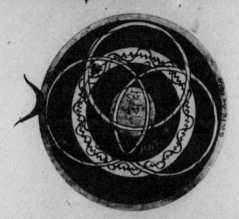

lxvij.   IS DE STELLA MARTIS ·

Astrologis ferunt stellas quas planetas uocant · Impeditissignis suum ita pagere cursum
Martion dini belli ee dicunt · Ce marte appellatu · quia p uiros pugnatur ut sit
marsmas· Licet et inusium genera consuetudinum sicut set taru ubi et femine et
uiris impugna eunt · Amazonu ubi sole femine · Romanoru aliaruq̄ gentiu
ubi soli mares · Le marte quasi effec morem mortiu · Ham a marte mors nuncu
patur huneda rerum dicunt quia belli gerentib; incertus est · Qd uo nudo
pectore stat ut bello sequisq̄ sine formidine cordis obicat · Mars aut apud gy
cos gnudius dr̄ eo qd in bello gradu inferunt qui pugnant aut qd impigt gn
Stella martis e quia alii herculis dixerunt   it yorxas   Iouin
ueneris sequens stella hae ut erat hostines ait de causa · qd uulcanus eu uxore
ueneram duxisset et appt eius obseruantia marti copia nfierct · ut nihil aliud
assequi uidetur nisi sua stella ueneris sidus psequi · auenere impetrauit ;
itaq̄ eu uehenti amor eum incende ret significans e facto stellam · phy riona

Figure Fifteen

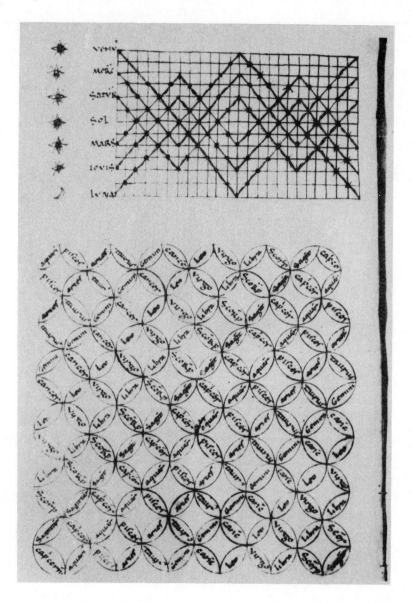

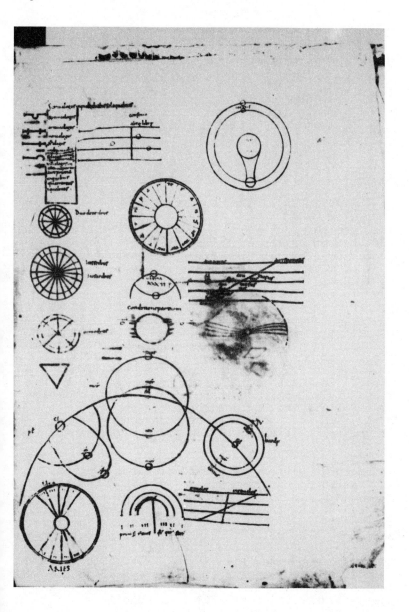

Chapter Twelve

PLINY AND RENAISSANCE MEDICINE

R K French

The previous chapters of this book have discussed
the greater or lesser fallibility of Pliny as a re-
porter of the natural world. The broadly critical
tone of much Plinian scholarship of the last two or
so centuries takes its origin from the renaissance
and in particular in Leoniceno´s attack on Pliny in
the last years of the fifteenth century. In an age
characterised by attempts to recover, restore and
above all admire the texts of classical civilisation, such an attack would seem to pose something
of a problem. The historian of medicine might see
in the episode similarities to Vesalius´ later
attack on Galen, so often seen as the first
successful attempt of the nascent science of anatomy
to throw off the burden of ancient authority.(1)
What then was the nature of Leoniceno´s attack: was
it the first stirring of a medical renaissance?

The Natural History in the renaissance
As Eastwood has explained in his chapter on
astronomy, parts of the **Natural History** were excerpted during the middle ages and used as texts
devoted to a single topic. This is true also of
medicine and related topics, for herbals were
constructed from parts of the work, and the
so-called **Medicina Plinii** is a collection of
remedies for diseases listed from head to toe. (It
was compiled perhaps as early as the fourth century;
there is an edition of 1509 from Rome.)
    The **Natural History** itself was one of the more
popular texts of the renaissance, with 15 **incunabula**
editions and at least 43 editions during the
sixteenth century.(2) What interested the
physicians of the time were Pliny´s accounts of
various kinds of **materia medica**, particularly those

described above in the chapters by Scarborough and Morton. Additionally, those interested in the structure of the human body found Pliny a useful source for good Latin terms for structures often also, and confusingly, known by their Arabic and Greek names. The problem of nomenclature was acute: in expounding the medical uses of herbs Greek medical texts generally gave only very brief physical descriptions of the plant, insufficient for the physician, particularly those of Northern countries distant from Greece, to identify it with plants growing in the field. Nor could he be sure which plant names in his own vernacular matched those in Greek, Latin and Arabic. William Turner found almost total ignorance about medicinal plants among the Cambridge physicians of the 1520s and could find illumination about the plants in Pliny (he owned the edition prefaced by Erasmus) only in Italy, Germany and Switzerland, with the new naturalists and philologists.(3) The **Natural History** was no doubt useful here, for Pliny was thought of as near enough a contemporary of, and familiar with the Greek sources, which he conveniently listed. His Latin terminology could be taken as authoritative translations of the Greek, and as an Italian he was no doubt familiar with the local and presumably near-Greek flora.

So when Leoniceno opened his attack on Pliny, he was not criticising a minor or obscure author. It must have been seen as a serious matter in the scholarly world. This chapter will sketch out the course of the dispute that arose from Leoniceno´s attack and then discuss some of the reasons behind it and the attitudes taken by a number of different people in two areas of medicine: it is not intended to give a bibliographically complete account.

The argument(4) began with an exchange of letters between Leoniceno and Politian. Leoniceno had expressed some criticism of Avicenna and some medieval authors, and he had also accused Pliny of confusing the Greek herb **kisthos** with ivy, **cissos**. Politian replied in January 1491, concurring in the criticism of Avicenna, but chiding Leoniceno for attacking so notable an author as Pliny on such slight grounds. This prompted Leoniceno to publish in the following year a detailed criticism, **De Plinii et Aliorum in Medicina Erroribus,** concerned primarily with herbs and fruits in medical use.

This pamphlet brought two very swift replies. The first was the **Castigationes Plinianae,** begun by Hermolao Barbaro some time before, but now modified to meet some of Leoniceno's critcisms.(5) The second was a much more direct response, the **Pliniana Defensio**, written expressly to refute Leoniceno by the lawyer Pandolfo Collenuccio.(6) While it was still in press he read Barbaro's commentary, and the two authors added a joint appendix to the **Defensio**. Leoniceno sought to justify himself in a letter to Barbaro who, however, died before Leoniceno had finished it. It became a second 'book' of the **De Erroribus** in the edition of 1509.(7) The third book of that edition is another letter, to Franciscus Tottus, in which Leoniceno rather half-heartedly returns to the attack on Pliny on the occasion of the discovery of a new manuscript. The 1509 edition is completed by two more letters of Leoniceno, the first of which had originally been published in 1498 and in which Leoniceno developed an earlier critique about the identity of the vipers used anciently in making theriac with the **marassi** of fifteenth-century Italy.

At first sight the dispute appears to have been a technical wrangle about the identification of the components of medicines. This had its important practical side, and indeed Leoniceno's expressed reason for attacking Pliny was lest the great powers of medicinal substances be wrongly used and jeopardise life (the point was made again by the editor of the collected tracts of 1509). But inevitably the practical business of identifying plants became, in the ensuing dispute, a question of terminology within the literature; just as (Leoniceno maintained) Pliny had confused **kisthos** with **cissos** because of the similarity of the words (rather than of the plants).

But behind the technical detail of the dispute there lay Leoniceno's perception of the **kind** of author Pliny was. Likewise Leoniceno's opponents attacked him and defended Pliny not only on technical detail but for the motives behind and the nature of his attack on Pliny. This will be examined in more detail below, but we can note briefly here that as a Hellenist, Leoniceno saw Pliny's work as sometimes clumsily derived from the superior Greek originals of Theophrastos and Dioscorides. It was in seeking pristine medicine that Leoniceno attacked Avicenna at the beginning of the dispute and the medievals Gentile da Foligno and Mondino dei Luzzi at other times. For Leoniceno they were all Latin interpreters. Similar reasons led him to attack Celsus, a figure in a position similar to that of Pliny in being an isolated Latin writer borrowing his material from a large number of older and more authoritative Greek sources. The belief in the superiority of the Greeks informed Leoniceno´s attitude to medicine. For example, in reporting the disputations at Ferrara about **morbus gallicus** he went against the widely held belief that it was a new disease by insisting that whatever its causes were, they were natural (it was not divine punishment); and it was unthinkable that the same natural causes had not operated in ancient times as recently. It was equally unthinkable that the ancients had not perceived these and the resultant disease, and the reason we cannot read about the **morbus gallicus** in the ancients is simply that this is a vulgar name and the ancients called it something else, since forgotten. Once again, continues Leoniceno, Pliny is wrong, this time in asserting that new diseases had come to Italy in his own time: it was only, said Leoniceno, that there were not then enough skilled Greek doctors in Italy to recognise such things as being known to the great Greek authors; and the forgotten diseases were held to be new.(8)

Humanism
These are the bones of the controversy initiated by Leoniceno´s attack on Pliny. To illuminate some of the reasons behind the actions of Leoniceno and his opponents and followers, I want to maintain a distinction that is not always made, particularly by historians of medicine. This is the distinction between Hellenists and humanists. In maintaining this I want to argue that the usual polarity between ´scholastic´ and ´humanist´ medicine is a false one, and that the relationship of medicine to such categories is often misunderstood.

The Italian **studia** in the early period of their existence taught vocational subjects including medicine and law. An essential part of these studies was the vocational use of rhetoric in a preparation for employment in affairs of government, politics and trade: the writing of letters and speeches, the **ars dictaminis**, taught by the **dictatores**. In general in the Italian middle ages no great use was found for the classical authors, and there were few classical models for these literary exercises. No commentary on a Latin poet or prose writer existed before the second half of the thirteenth century.(9)

In contrast in France there had been an interest in classical texts since the ninth century, and there was a considerable amount of classical study at Chartres and Orleans in the twelfth century. Italian humanism of the later thirteenth century was partly an importation from France. The new ´humanist´ teachers in the **studia** were the professional successors of the **dictatores**: they held chairs of grammar and rhetoric (later also of prose and poetry) but unlike their medieval antecedents they taught from examples of classical Latin prose.

The important point of this for us is that humanism was not only a style of teaching, but a style of teaching **certain subjects**, like the **ars dictaminis** it succeeded. Kristeller(10) points out that the term ´humanist´ arose, perhaps from student slang, to match the term ´canonist´ (for the teacher of Canon Law), ´jurist´, ´legist´ and ´artist´. The older term for the group of subjects taught by the humanist was **studia humanitatis,** embodying the idea of a liberal education, borrowed from Cicero and Gellius. By the first half of the fifteenth century the **studia humanitatis** included grammar, rhetoric, history, poetry and moral philosophy. The humanists used classical Latin models extensively in teaching (with a little Greek) and produced great quantities of commentary employing their historical and philological techniques of criticism. A number of subjects were excluded, then, from the **studia humanitatis:** medicine, law, natural philosophy, logic, metaphysics, astronomy and theology. Yet in the second half of the thirteenth century some of these subjects were also undergoing great changes. In particular (for our purposes) the reception of the natural-philosophical works of Aristotle opened up a large new field of study. In medicine the Latin translation of the **Canon** of Avicenna quickly came to occupy a major part of the teaching course. Both of these important sources lay outside the **studia humanitatis,** not only by reason of their subject matter, but because neither had been written in Latin and so could not provide models of Latin prose. The same is true of other, earlier translations from the Greek, which were concerned with technical subjects like medicine, astronomy and mathematics. An important text in this group is Galen´s **De Usu Partium,** translated by Niccolo of Reggio in 1317, but earlier available in Greek to such scholars as Peter of Abano (who claims to have read it).

The natural-philosophical works of Aristotle, the **libri naturales,** were taught in the **studia** principally by medical men. Indeed, the peripatetic world picture came to be the basis of the theory of medicine (partly through Galen´s interpretation of Aristotle). Now, these processes, the assimilation and teaching of Aristotle´s physical works, involved close examination of the text and the locating of each of them in its proper place in natural philosophy as a whole. All this was carried on by means of Aristotelian modes of argument and presentation: in a word, scholastically. Scholasticism had (like humanism) arrived in the Italian **studia** after their early vocational period (and perhaps also from France). To remember that in Italy scholasticism was no older than humanism and was principally distinguished from it by the subjects it taught will help the medical historian (in particular) to divest the term ´scholasticism´ of some of the pejorative sense that still clings to it (as a result of the humanists´ renaissance battle against it).

What relationship, then, did medicine have to humanism? It seems likely that in the century or so between the death of the great scholastic commentator Gentile da Foligno and the birth of Leoniceno (in 1428) medicine had absorbed much of what humanism was able to offer to it. In the Italian universities where arts and medicine were taught side by side it would be surprising if the humanists´ interest in classical texts, and a philological and historical scholarship had not had some impact on medicine. Certainly by the last decades of the fifteenth century there is abundant evidence(11) that medicine shared the three great characteristics of humanism: it had its classical texts; philological scholarship strove to reconstruct the meaning of words as used by the authors in their own time; and a strong historical sense helped the medical men see the products of the past as formed by historical circumstances.

Indeed, the writings of medical men from Gentile to Berengario da Carpi (who as we shall see engaged in controversy with Leoniceno) betray little antagonism between humanists and physicians. Much clearer is the clash of interests and purposes of the natural philosophers and the medical men, for there were many natural questions to which they could not give similar replies. Since medical theory was based on Aristotelian physical principles the philosophers felt themselves qualified to pronounce on medical questions, at least in giving true causes; the medical men could not deny that medicine was part of philosophy, because Galen had largely Aristotelianised medicine; but nevertheless the medical men held that in at least two main areas of medicine - anatomy and the action of drugs - only sensory experience, not logical demonstration, could lead to true knowledge. The attempt of the medical men to establish a professional preserve against the philosophers - a **via medicorum** - is hinted at in Gentile(12) and clear in Berengario. These points - the subject-based nature of humanism, the partly humanised medicine of the later fifteenth century, the uneasy relationship of philosophy with both medicine and humanism (which did not come to grips with it until this time)(13) - set in greater contrast the last of the groups we have to discuss, the Hellenists.

Hellenism

The people whom history presents to us as taking part in purely intellectual disputes must often have been moved by more immediate and compelling motives. Such was the case with the Hellenists of the fifteenth century. Many of their actions can be explained on the basis of their political circumstances. For many of them were not merely Hellenistic in outlook, but were, in person, Byzantine Greeks. For our purposes the story begins shortly before the introduction of humanism and scholasticism into the Italian **studia.** In 1204 the fourth crusade sacked Constantinople and a ´Latin Empire´ was set up over Byzantium. The Italian cities, and particularly Venice, made great use of the new commercial markets that appeared, and Venice for a long time retained colonial possessions in the East. For the Byzantines the affair generated a long-lasting hatred of the Latins, whom they saw as culturally inferior ´barbarians .

Constantinople was regained (in 1261) by Michael VIII Palaeologus, and there followed a ´Palaeologan renaissance´ which emphasised the Greek culture of the Empire at the expense of its Roman origins, for ´Roman´ was the designation of the Western barbarians. , But this was soon followed by increasing pressure from the Turks to the East. Byzantine emigrés even appeared in Italian cities. Finally the Byzantines were compelled to put to one side their hatred of the Latins in an attempt to save their state from extinction at the hands of the Turks. During the Council of Florence some 700 Greeks spent eighteen months in Italy in an attempt to unite the Christian churches of Rome and Byzantium and so secure Western military help for the defence of the East. This military aid failed to materialise and The City fell to the Turks in 1543.

But the Council of Florence had effects on the relationship of East and West that are of interest to us. First, the Italians were duly impressed by the culture and philosophy (particularly Platonism) of the Byzantines. Second, despite their lingering belief in the cultural inferiority of the West, many of the Byzantines were impressed with the vigour and wealth of the Italian cities. Some of them stayed in Italy after the Council and others returned, to be followed by more of their compatriots up to and after 1543. Venice´s Greek possessions survived the fall of Constantinople and so in addition to the intellectual attractions of the new philosophy from the East, there were sound practical reasons of colonial administration and commerce for the younger sons of the nobility to be taught Greek.

The history of the teaching of Greek in the West also reflects these political and commercial circumstances. Although the first Western chair of Greek was established in Florence in 1361 on the urging of Boccaccio, it is argued convincingly by Geanokoplos(14) that the real beginning of Greek studies dates back to Chrysoloras, who had been at the Council of Florence and who took the chair of Greek at Florence. One of the advantages of setting up chairs of Greek is said to have been the practical one of of recovering and practising Greek medicine;(15) but it is likely that the political and mercantile advantages were at least as clear. This is illustrated by the chair at Venice, held from 1504 to 1506 by Nicolaus Leonicus, who had studied with Chalcondyles, a pupil of Chrysoloras: this was not a university post but was attached to the Ducal Chancery School for the training of civil servants for the Venetian Greek possessions.

For their part, the teachers of Greek used their positions to express their nationalism and sense of cultural superiority. Before the fall of Constantinople these teachers spoke from their chairs of the necessity of saving their beleaguered countrymen; and afterwards, they called for help for emigres or even for the establishment of a new Greek state. Thus Bessarion, a veteran of the Council of Florence and now a cardinal, opened his house to other Greeks and called for the formation of a collection of Greek manuscripts for the preservation of Greek learning. Pletho (whose Platonism had been so interesting) had an ideal of a new Hellenic state; it was to have been a new Greece, not a new Byzantium (which had been originally Roman). The cultural superiority felt by the Greeks is clear in Chalcondyles´ introductory address to the university of Padua in 1463. Despite some mock humility he makes it clear to his audience - one might say offensively clear - that he believed that everything of value in Latin culture had been given by the Greeks.(16)

These then, were the Hellenists, and it is helpful if we see them, at the time when Leoniceno was attacking Pliny, as a distinct group with certain things in common. They were certainly seen as a group by the medical men, for example by Berengario and Gabriel de Zerbi.(17) If we look at medical writing at the time we can distinguish **academic** medical men, who taught in the universities and who had absorbed much of both the scholasticism (they called themselves **scholastici**) and the Latin humanism of the universities, from the Hellenists, who were largely outside the universities, whose prime concern was being Greek, and who in a number of cases (like Chalcondyles)(18) began with only a rudimentary knowledge of Latin and its culture.

Before looking in more detail at the fate of Pliny at the hands of the medical men, we should note one more characteristic of Hellenism. In emphasising the Greek origin of all good things, the Hellenists were obliged to step back, over the Roman foundation of the Eastern Empire, to Athenian philosophy and science. Ancient Greek science was the **prisca scientia**, borrowed but not properly used by the Romans, and stolen by the Arabs. Like the Western humanists, the Hellenists´ business was to recover the culture of antiquity, but unlike their Western counterparts they saw Latin only as a language of transmission, and not a language in which the **prisca scientia** resided: it was the language of the commentators. But Greek philosophy antedated Christ and was pagan; in emphasising their Greekness, the Byzantine emigrés sometimes seemed more Hellenist than Christian. It has been suggested that Pletho´s ideal of a state of Hellenes involved the disestablishment of the church.(19)

In a similar way the classical sources for Western humanism were largely non-Christian. Hellenists and humanists agreed broadly on the range of topics proper to the liberal education they sought to establish on classical models, excluding medicine and the other topics listed above. In contrast the academic medical man saw himself in a tradition of scholarship that concerned itself with topics (like medicine itself), authors (like Avi-

cenna and the commentators) and a period of history
(the high middle ages) most of which were excluded
in the search for a **prisca scientia**. The academic
medical teacher of the partly humanised medicine of
the time used the humanists' tools of philological
and historical criticism on the texts of the
medieval Latin and Arabic writers; and as a
technical writer, he even viewed the Hellenists with
suspicion in choosing the elegance of antique
letters as an easier option than the rigours of
academic scholarship.(20)

It can also be argued that the academic medical
man was more concerned with the Christian middle
ages than were humanists and Hellenists. The
arrival of the Aristotelian **libri naturales** and of
Avicenna's **Canon** produced generations of commentary
which was still essential reading for medicine in
the early sixteenth century. Of course, Aristotle's natural science had to be modified, for
example by dropping his claim that the world was
eternal and by introducing a (rather Platonic)
creator God, who better fitted Christian doctrine.
God could then be represented as creating the world
using principles of Aristotelian physics.(21)
Avicenna' Platonised Aristotelianism was also
acceptable.(22)

In certain circumstances the Christian objection
to Aristotle's cosmology could be paramount. This
happened at the reformationist university of
Wittenberg in the early sixteenth century. In
general, the arts course in European universities
was almost entirely Aristotelian, beginning with the
universal arts of argument and proceeding to their
application in the sciences. Physics was the first
science, the general study of the principles of
physical change, and the student then proceeded
through more complex sciences of actual physical
motion, passing through meteorology to Aristotle's
De Anima or perhaps **De Animalibus**. The internal
coherence of this scheme was great, and the early
parts of the course were necessary stepping stones
for the later; and after the final Aristotelian
works the student was ready to proceed naturally to
the still more complex study of medicine. Indeed,
so rational and widely adopted was this scheme that
it must surely have been a decision of some moment
to abandon Aristotle in favour of Pliny for the
early arts course at Wittenberg.

The reasons behind this decision were surely to
do with Wittenberg's role in reformist thinking.
Passionate concern with the principles of religion

may be assumed among those attempting to reform the church, and such concern would highlight the ways in which the usual Aristotelian teachings of the universities were not strictly Christian. We saw in the previous chapter that astronomy did not necessarily provide evidence for Nauert´s thesis about the protestant background to these changes at Wittenberg, but there is abundant evidence that the **Natural History** was used by Milich at Wittenberg because it could provide a basis for science that was more agreeable to the principles of religion than Aristotle´s physical works.(23) Milich´s argument is that the **Natural History** is about nature, and a knowledge of nature leads to a knowledge (**agnitio**) of God. With Pliny´s help, souls are led to true religion, which is defended by the stout walls of the **Natural History.** Milich can use both a text from St Paul (that divinity is so close in nature that we can almost touch it) and a Platonism (philosophers are ´just´ in being led by nature to God) to maintain an argument from design to Designer.

Milich´s purpose is to use book II of the **Natural History** to replace the physics of Aristotle in teaching his young students. This was a considerable undertaking, given that in the usual university course the whole of the later curriculum rested on the basis learned at this stage: what is there in Pliny comparable in scope and logical rigour to the basis and development of Aristotle´s physics? Answers had to be found, because it was quite clear to Milich that Aristotle´s physics was unsuited to describe the actions of divine will in creating and maintaining the world. The fundamental argument against Aristotle was that the world was not, in fact, eternal, as Aristotle maintained, but created: Milich spends much space on this, rehearsing the arguments in an academic way. Chief amongst them is that efficient causality cannot stretch back in an infinite chain, but must at length depend on a First Cause. Notice that in trying to sweep aside Aristotelian physics, Milich cannot however do without causality, efficient, formal, material and final: God is the efficient creator of the world and His purposes for it are its final cause; material and formal causes Milich finds in Plato. Further inadequacies of Aristotelian physics for Milich is the peripatetic denial of a Creative power, **nihil ex nihilo fit,** and the necessary relation between cause and effect: fire is obliged to burn. No, asserts Milich, a Creator can produce matter from nothing as He can equally suspend the operation of the elements and their qualities.

The basis of a **real** physics for Milich is God´s creation of the world as a habitat for man. Final causality involves the providential design and usefulness of nature for man, by means of understanding which man comes to know God. The sequence of efficient causes from the First Cause to effect is Necessary Causality in Milich´s theological physics, and he finds it necessary to defend his real physics against the pretended physics of the Epicureans, who would deny efficient and final causality. Milich´s causes operate firstly on ´nature´ in the general sense of the laws by which the world operates, and secondly on **particularis natura** of bodies, or their temperament, derived from their elements, qualities and faculties. For all his metaphysical re-orient-ation, Milich is back here to the standard components and powers of the natural world, by which the physician for example, would understand such Naturals as innate heat.

In short, Pliny´s admiration for nature as the habitat of man was much more sympathetic to Milich than Aristotle´s world of ´philosophical´ causes. While there were religious reasons for this, this is not the place to enter into a discussion of whether the Wittenberg lectures on the **Natural History** was due to the Protestant view of things. But it should be noted that the people whom history normally calls the new natural historians of the sixteenth century (such as Gesner, Turner, Belon, Rondelet) all read the **Natural History** in the same way as Milich, and conceived it to be their duty to come to a knowledge of God through a study of His creation, the natural world. And in fact few of them were untouched by a protestant theology. Milich´s concern for the education of his young students is not only that they should appreciate the religious utility of the **Natural History** but that they should do so in the usual academic way: his commentary is based on the common practice of scholars, the "scholastic disputation".(24)

Leoniceno and Pliny
It is against this background of Hellenists, human-ists and academics that we can profitably ask why it should have been Leoniceno who attacked Pliny. The answer seems to lie in his personal interpretation of the Hellenists´ message.

Nicolo Leniceno began his Greek studies at the
age of fifteen with a relative, Ognibene Leoniceno.
Through Ognibene, Nicolo was not far removed in a
line of teachers and pupils from Chrysoloras, the
Byzantine emigre and first great teacher of Greek in
the West. We have no direct evidence of the transmission to Leoniceno of the passion with which the
Byzantine emigres argued for the cultural superiority of their civilisation, but the circumstances
are suggestive. Chrysoloras had been at the
Council of Florence as one of the Byzantine delegation, and the urgent political needs of
Constantinople must have quickened his sense of
cultural identity and of history. The Byzantine
delegation first met at Ferrara, Leoniceno´s town,
when he was ten, in 1438, and it difficult to
believe that the significance of the events was not
explained to him. When he began his Greek studies
five years later, it was still ten years before the
Emperor and his subjects chose to die at the hands
of the Turks in the City where their ancestors had
lived for eleven centuries rather than abandon
it.(25)   (In that year too, the Byzantine Theodore
Gaza began to teach at the reformed **studium** of
Ferrara.) When Constantinople fell, Leoniceno was
undoubtedly the master of the language, and it is
not difficult to believe that the loss of a
civilisation deepened in the minds of the Hellenists
- including Leoniceno - an awareness of its merits
and the importance of its records. Leoniceno took
his doctorate in that year, 1453 (and performed the
usual academic exercises alternately in Greek and
Latin).
Apart from these hints as to Leoniceno´s
Hellenism, we should note that he combined it with
medicine (and mathematics and philosophy). It was
not perhaps the most natural of combinations. Just
as Western humanism did not in its origins and use
naturally extend to the sciences or philosophy, so
the Palaeologan Greek renaissance was not primarily
an affair of the technical subjects. Of course,
the authorities of the **prisca scientia** of the Greeks
included Hippocrates and Galen, and so the
Byzantines like Chalcondyles could still when faced
with medicine maintain the superiority of their
culture.
In general however, the humanists and Hellenists
avoided the technical subjects of medicine and
philosophy. The Hellenists often identified themselves as writing with eloquence´, by which they
generally meant ´in Greek´. Those who knew Latin

only, they said, 'stuttered' in their expression, and nowhere more so than in expressing the matter of the technical subjects. The practitioners of these subjects sometimes retorted that the Hellenists were mere grammarians, unable to cope with the rigours and discipline of academic subjects like medicine and philosophy.(26) It seems to have been the case that many of the Hellenists (rather than humanists) were outside the universities and university based professions, so that 'academic' can be opposed to 'Hellenist' in this way too.(27)

Leoniceno, however, was not content that the Hellenists should make only eloquence and good letters their proper field of study and asserted that the time had come for the Hellenists to tackle the technical subjects of medicine and philosophy.(28) He taught his own students (such as Tottus) both medicine and the liberal arts,(29) and no doubt saw himself as a model philosopher and physician who like his great antecedents Plato and Galen conducted his business in Greek. He stigmatised his enemies as being without a knowledge of medicine or Greek and referred even to fellow Hellenists as 'mere grammarians' if they did not have medical knowledge. Merula was a mere grammarian (30) in Leoniceno's view because he worked on Latin rather than Greek texts (and was the teacher of Leoniceno's enemy, the Patron of Pliny, whom we shall meet below). Even so Greek a person as Theodore Gaza, the translator of the Aristotelian zoological works, who had been appointed to the **studium** of Ferrara at about the time when Leoniceno bagan his study of Greek, was a **grammista**.(31)

In urging the Hellenists to extend their empire over more educational fields Leoniceno was giving a new direction to what was already a vigorous campaign for Hellenistic studies in education both within and outside the universities. The handful of people who did in fact combine Hellenism with medicine or philosophy felt themselves even more than the Hellenists as a whole to be fighting a crusade against an inferior but vastly more numerous enemy. Their self-image was never far in fact from the last defenders of Constantinople. Menochius wrote to Leoniceno in 1503, urging him to continue with his translations of Galen, fearing that if they were to come to an end "...just as the barbarians of Turkey destroyed a great part of Greece, so the medical barbarians of our age will confound confusion and will lacerate Galen with their inept translations".(32) Leoniceno spoke of medical and philosophical Hellenists as 'the few' who were 'proven in war'; Tricaelius presented Leoniceno to the reader in military terms as the victor over rustic barbarity(33): in the Hellenists' rhetoric the barbarian Turks were readily equated with the barbarians who had overrun the Roman Empire and imposed a medieval gloom over Italy's literary studies, and with the Arabs, who had stolen from the Greeks the claim to have discovered medicine.(34) Leoniceno and his circle addressed each other in their letters and dedications as those alone who had the ability and, above all, duty to combine eloquence with medicine and philosophy and so free Italy (or the republic of letters, or medicine or philosophy) from the barbarians (or the rustics or the monks).

There are many ways, then, in which Pliny fell short of Leoniceno's ideal. He did not write in Greek; indeed he was not Greek. Certainly he read Greek, but that only meant that the bulk of the **Natural History** was taken from Greek sources. And these sources, in the topics in which Leoniceno chose to attack Pliny, were the older and more authoritative Dioscorides and Theophrastos.(35) Leoniceno maintained that unlike Theophrastos, Pliny had no adequate scientific method, and that unlike Dioscorides Pliny dealt with words rather than things. Nor did Pliny have any special knowledge of philosophy and medicine, essential components of Leoniceno's ideal scholarship. At best, Leoniceno said, Pliny was a grammarian or orator(36) and later claimed that most people read Pliny for his vocabulary and style rather than for the technical content.(37) And it was in fact only the Latin eloquence of Pliny, whom Leoniceno calls with Hellenistic condescension Pliny the Latin Man,(38) that saved him from coming into the same category as the barbarians. Pliny, said Leoniceno, was 'in the same game , **in eandem aleam,** as the barbarians.(39)

In reading Pliny, Leoniceno as a Hellenist physician could hardly have found sympathetic Pliny's attack on Greek doctors and medicine (described in Nutton's chapter); and Pliny's preference for old Roman rustic simplicity may have seemed to Leoniceno another area in which Pliny was dangerously close to being barbarian. Perhaps too, Pliny's jaundiced view of 'Greek vanity' (compare the **vanitas Graecae** of Greek astronomy, p.187 above) recalled to Leoniceno contemporary scoffing, by the academics, at the elegant **graecitas** of his group of Hellenists. Leoniceno's aim was to defend Hellenism as much as to belittle Pliny.

In Leoniceno´s view these shortcomings of Pliny became manifest in the **Natural History** in two principal ways. The root of the problem was Pliny´s lack of philosophy which, said Leoniceno, was the cause of Pliny not practising an adequate scientific method. The results of this were several categories of errors and confusion of fact of the text of the **Natural History: vicinitas nominum,** confusion of similar names of dissimilar things (**cissos** and **kisthos** being the **locus classicus**); **aequivocatio,** ambiguity, were a single name might suggest to the unwary compiler more than one thing; a single thing with more than one name, each taken by the hasty encyclopedist to represent another thing; and similarity between things taken rashly as identity.(40)

The method that would have avoided such errors, and in which Pliny was deficient was, said Leoniceno, the best scientific method of the Greeks, based on the superiority of sense observation over reason, and of knowledge over (written) opinion. In practising the method (said Leoniceno), large numbers of observations were brought together to form a single statement, which was then divided in accordance with its now apparent natural articulations to reveal truth. According to Leoniceno the best practitioner of the method had been Socrates, but it was Aristotle who had used it to the greatest extent (with some difficulty however in the zoological works). Likewise Theophrastos, the philosophical successor of Aristotle and the literary source for Pliny, was a skilful practitioner of the method, unrecognised by Pliny. And of course, Galen likewise earned Leoniceno´s praise, for in using the method he was philosophical, medical and Greek.

Leoniceno demonstrated his ability to live up to his own ideal as a philosophical, medical and eloquent scholar in much of what he wrote; and particularly in his treatise(41) on the ´three doctrines´ announced so tersely by Galen at the beginning of the **Ars Parva** that they were subject thereafter to much commentary. Here Leoniceno does indeed deal with the technical matter of logical forms and arguments in a very lucid way. In doing so he is remedying what he saw as defects in the academic treatment, particularly that of the commentaries. He may have been thinking of the commentary on this text by Gentile da Foligno, whom Leoniceno often describes as one of most important of the **iuniores,** the recent (that is, medieval) and

barbaric commentators. Whether or not he had
Gentile in mind, the latter's commentary(42) is
characteristically medieval, and his discussion is
abbreviated, unrelieved by example, highly technical
and less considerate to the reader than Leoniceno's.
But of course Gentile expected his readers to be
entirely familiar with the kinds of technical
detail and the presentation, and he can address them
with all scholarly rigour and comprehensiveness.
Leoniceno on the other hand knew that his Hellenists
were on the whole nervous of the technicalities of
the scholasticism of Gentile's day and of the
academic treatment of their own, and he has to lead
them by the hand, lighting the way with eloquence
and providing easy examples to show how the methods
worked.

The defence of Pliny
From what has been said above it can be seen that in
attacking Pliny Leoniceno was putting into operation
his programme for a Hellenistic invasion of academic
territory. He could attack Pliny on medical matters (the identification of simples), philosophical
(scientific method) and linguistic (confusion of
terms). We need not be surprised that the
academics (for example Berengario) resented this
invasion, and those who (like Collenuccio) defended
Pliny did so as Leoniceno had attacked him, that is,
at a level deeper than the factual accuracy of
details: that Pliny was a kind of author worthy of
defence. And even more relevant to our interest is
that the academics' reply was determined partly by
the kind of author they thought Leoniceno was.
  This is clear from Collenuccio's defence of Pliny
against Leoniceno.(43) Even in its external
appearance - black letter, double columns - Collenuccio's text is clearly a product of the schools,
resembling for example the nearly contemporary
anatomy text of Zerbi(44) and the weighty folios of
commentary on Avicenna printed in Venice for the
Italian medical schools.(45) Leoniceno and the
Hellenists were always careful to have their words
printed in a plain roman uncial or in italic,
without columns. Collenuccio's academic subject
was law, which like medicine, could only accept so
much humanism and which by its very nature (it was
Roman law) could be less readily Hellenised than
medicine (where the Greek authors were important).
Collenuccio does not pretend, he says, to study
'eloquence' (Greek) or to have a detailed knowledge
of medicine: he modestly says he is versed "only in

the most illiterate of letters, since I am a lawyer".(46) Indeed, Collenuccio groups himself with the **scholastici**(47) who urged him to reply to Leoniceno, and like these academics he suspected Leoniceno of an excessive zeal for things Greek.

It was as an academic that Collenuccio resented Leoniceno's attempt to extend the disciplinary boundaries of Hellenism. He recognised these boundaries, and knew that he would himself risk criticism in straying too far from his own professional field of law into those where he had no special knowledge, "using his scythe in a foreign field" as he expressed it. But the Hellenists' attack could be resisted by taking the initiative and altering the boundaries in favour of the academic and technical subjects. Collenuccio argued that one philosophy should embrace all the "arts, disciplines and letters". As a lawyer speaking from the stronghold of technical argument ("dialectic and philosophy"), he saw 'eloquence' and 'letters' as something to be brought within the grasp of the academics.

Naturally Collenuccio arranges his defence of Pliny in a legalistic way, first appointing himself the "patron" or advocate of Pliny. The main force of his 'case for the defence' of Pliny - and it was echoed by others who disagreed with Leoniceno - was not so much that the attack was incorrect in the disputed details, but that it was **unfair**. We should here pause to remind ourselves that the principal model for conducting an argument and discovering truth was the disputation, the structure of which was so thoroughly absorbed by university-trained men that it perhaps even unconsciously determined their conception of what constituted proper discussion and proof. The essence of the disputation, and of the legal procedures with which Collenuccio was familiar in court, was that the argument went from one side to the other until a resolution was reached. But as Collenuccio reminded his readers, Pliny had been dead for more than a thousand years, and unlike the disputant or litigant was **not there to defend himself**. What was involved was not simple injustice, but that the argument could not reach a satisfactory conclusion, and truth discovered, without one party being present to make **distinctiones** and to offer, and resolve the **quaestiones** and **dubia** that were the necessary stages of the **disputatio**. "I wish that Gentile were alive, so that he could answer my **quaestiones** and **dubia**!" exclaimed Leoniceno on an anatomical matter.(48) But Gentile, like Pliny, being dead, the argument could only proceed with an adopted Patron to speak for the absent disputant: Benedetti in the anatomical matter, as we shall see, and Collenuccio in the present. Collenuccio's models are Cicero and Crassus, **prisci** but of course Roman advocates. The unfairness of Leoniceno's attack on Pliny is contrasted by Collenuccio with the strict legal procedures by which a fair judge always defends the rights of even an absent defendant to a fair trial. Moreover, adds Collenuccio, even in cases of confessed crime (even if Pliny sometimes **is** wrong) the defendant escapes punishment if he is sufficiently useful to, or a great ornament of, society. Pliny of course - and everyone who defended him did so partly on these grounds - was a great ornament of **Latin** culture, which Collenuccio and others called the **Res Latina**. (49)

The **res Latina** - the term seems to be a conscious reaction against the **Graecitas(50)** of the Hellenists - involved the academic subjects that Leoniceno was urging his fellow Hellenists to invade. In the later dispute for and against Pliny, Leoniceno, his followers and the academics further illustrate the battle lines that were drawn up. Leoniceno's anonymous pupil and defender, the 'Roman Physician', felt he had scored a particular triumph when, in line with Leoniceno's programme, he maintained his defence of Leoniceno's account of the three doctrines of the **Ars Parva**, not only before the modern **scholastici**, but in the terms of their teachers who, the Roman Physician confesses, were very learned men.(51) These teachers included the 'old expositor' of Avicenna, Gentile da Foligno, the 'new expositor', Jacobus de Partibus, and 'the Monk' or Turrisanus, the **plusquam commentator** on the **Ars Parva**.(52)

These medieval figures were very much part of the tradition of Latin scholarship of the academics, and their commentaries were well known and were printed in the early sixteenth century. Broadly, the Hellenists wished to avoid the commentators; and in the scorn with which they used the words the Monk' and 'monkish' in general(53), there is a hint of the difference between the pre-Christian philosophy of their intellectual heroes and the superstitious religiosity they saw in the medieval Latins.

That the 'Roman Physician' found a minor triumph in matching the learning of the medieval scholastics reflects a common feeling among the humanists and Hellenists that the medieval commentary was not only

inelegant but formidable. The medieval medical
commentary had a very rigid and complex form and its
formal **expositio** section, at its most elaborate,
could reduce a bulky text like that of the **Canon** of
Avicenna to a series of statements so short that
each could be treated as the proposition of a
syllogism and so exposed to the whole apparatus of
argument. The **expositor** did this without summary
or omission of the text, and his purpose was to
expose the flow of author´s argument in its entire-
ty.(54) Medieval medical commentaries could be
twenty times as long as the text, and to one not
brought up on them as a staple diet of education it
was (and is) a difficult business to get ´inside´
them. The humanists and Hellenists did not want to
do so. Leoniceno, mocking the successive quotations of an early stage of the **expositio** as
"...primam ibi, secundam ibi, tertiam ibi..." argued
that the technique (by which the commentator divided
his text) rather broke up the text and argument than
allowing it to speak for itself. Turrisanus´ **expositio** similarly offended Leoniceno´s Roman pupil,
and Leoniceno called it "all **ostensio** and no
use".(55) But the medical men in the universities
still read the great commentaries and even, like
Berengario in Bologna, composed new works in commentary form. They found the **periti Elleni** elegant
but unwilling or unable to cope with the technical
matter and techniques of medicine and philosophy,
and dismissed them as grammarians: Leoniceno´s Roman
Pupil, attempting to be as medical and philosophical
as his teacher, felt the insult keenly. Barbaro,
who brought out an edition of the **Natural History**,
resented being dismissed by the scholastics of Padua
under their leader, the "Ape of Padua" (said
Barbaro) as one of the **insensissimi** Hellenists, who
were to be avoided.(56)

Pliny and the anatomists
It was observed at the beginning of this chapter
that the **Natural History** was, for the medical men of
the renaissance, a source not only of accounts of
items of **materia medica,** but of Latin anatomical
terminology. Leoniceno´s criticisms of Pliny were
as sharp in this field as in the other. But again,
not all anatomists were Hellenists of Leoniceno´s
kind; and the practice of dissection was common
enough to provide physical criteria for the resolution of disputes about terminology.
As he was advocating the Hellenisation of
medicine, Leoniceno resented the attempt of a medic-

al man to ´usurp´ the function of the literary man
in emending the text of, and in defending Pliny.
Leoniceno bitterly attacked the ´medical Patron of
Pliny´, whose duties, he claimed, should be to try
to understand the workings of nature, not to meddle
with texts: the Patron had not only published an
edition of the **Natural History,** but had written an
anatomical textbook embodying Plinian errors.
Leoniceno´s rhetoric of invective included the
device of denying the Patron the dignity of a name,
but from the quotations he makes of his oppon-
ent´s anatomical opinions, we can identify him as
Alessandro Benedetti.(57)

In using the terms **patronus** and **cliens,** Leoniceno
may be going beyond the legal use Collenuccio made
of the terms when he made himself Pliny´s **patronus,**
advocate. When Leoniceno called Benedetti
´Patron´, and when his Roman pupil used the term for
the modern Turrisanists, it was meant as an insult.
What they may have had in mind was the civil
patron-client relationship of ancient Rome, in which
the patron, generally rich or noble, fed and
protected the client, who may have been poor, a
non-citizen or a freed slave. The benefits to the
client were obvious; and the patron gained prestige
by the number of his clients, and could call upon
their services. But the relationship was binding,
and the patron was obliged to protect the client
irrespective of the merit of the client´s opinions;
nor could civil or legal patrons and clients sue
each other. Perhaps Leoniceno saw Pliny as a
client, deficient in true scholarship but
nevertheless necessarily defended by Benedetti, the
better informed patron, committed right or wrong to
his client.

Leoniceno differed from Benedetti principally on
the place to be given to Pliny and Latin culture in
the classical revival of the renaissance. This
difference is highlighted by the fact that in many
things Benedetti appears to have shared the ideals
of the Hellenists. He practised medicine in Greek
speaking areas for sixteen years, and knew the
language; he was a member of a circle of Hellenists
who supported and advised each other; and he
declared that the Greeks were the true founders of
all medicine. In dedicating his **Anatomice** to the
emperor Maximilian, the **humanissimus princeps,** he
promised to guide him through the terminology as
Pollux had guided Commodus: the learned Greek
scholar informing the powerful Roman emperor. In
this dedication, moreover, he declares that anatomy

is a purely Greek matter, and it is clear that he is rescuing a **prisca anatomia** from the clutches of later barbarians who claim the Greeks´ renown as the discoverers of medicine. He anticipates some criticism from the academics on this question, but dismisses it in advance, talking over their heads directly to his prince.

But apart from the obvious Hellenism of the **Anatomice**, Benedetti embraced also the **res Latina**. He believed (with Leoniceno) that eloquence should enter the technicalities of medicine, but for him eloquence included Latin, and the **externa lingua** used by those without eloquence was the vernacular, or barbaric Latin, or Arabic. Greek and Roman dignity were both pristine and both to be rescued. They differed, of course, for the Roman virtue of **gravitas** did not seize readily upon the subtleties of Greek theory, and so (Benedetti implies) medicine remained Greek. In Benedetti´s prefatory material in the **Natural History**, the **Anatomice** and his collection of medical aphorisms, we miss the urgent tone of the Byzantine writers, the sense of Greek identity. For Benedetti, Greece was purely classical and had long ceased to exist; but there **was** continuity with ancient Rome: not only did the educated in Pliny´s time speak the same language as the educated in Benedetti´s, but Pliny was Italian, even (Benedetti believed) from Benedetti´s own town, Verona. The title and the preface of Benedetti´s edition of the Natural History hail Pliny as ´our citizen´.

In practice Benedetti´s anatomical debt to Pliny was limited to terminology. Like Pliny, Benedetti listed the sources he was using and it is clear that the anatomical information, the description of structure, was taken from the Greek authors. The Latin authors are used - because Benedetti is writing in Latin - to supply good Latin names for the parts thus described. Most of the discussions that arose among the anatomists over Plinian terminology concerned the structures of the throat and the superficial parts of the face. There were also discussions on the oddities of animals and man as reported by Pliny, particularly where the **Natural History** differed from Aristotle´s zoological works.

In addition to the humanist Benedetti, we can profitably glance at Berengario da Carpi, whom Leon- iceno and Fallopio regarded as the greatest anatomist of the time.(58) Berengario was an academic: he had had an early surgical training from his father, a late humanist-Hellenist education

from Aldo Manuzio and then a formal medical training
in the university of Bologna. He had a taste for
the elegant and antique, but wrote with a
businesslike Latin that was closer to that of the
medieval surgeons than to Cicero. He used to great
effect the humanists' tools of philology and
historical criticism, but did so also on Arabic
texts. He admired Galen above all authors but was
immensely well read in Avicenna and the commentators, and indeed cast his own anatomical work in
the form of a commentary. He strove for a purity
of terminology, but accepted uncorrupted Arab words
as perfectly satisfactory technical terms. He
approved of the efforts of the "learned Hellenists"
(in whose number he held Leoniceno and his Roman
pupil) to establish a sound anatomical terminology,
but he attacked them for avoiding the difficult
business of true (academic) scholarship by hiding
behind the elegance of 'good letters'; he even took
the battle into Leoniceno's camp in accusing him of
mistranslating from the Greek. Berengario was, in
short, a good example of an academic practitioner of
the medicine of the early sixteenth century, which
had absorbed what it could of Western humanism and
which was felt by its practitioners to be distinct
from the newer Hellenism.(59)

Berengario's personal interpretation of anatomy
was to lay great stress on the role of sense
observation, and so his criterion for assessing the
written authorities was whether the authors had used
observation or had been simply **aggregatores**, putting
books together from other books. In this way too
Berengario saw himself and his science as distinct
from most of his scholastic forbears (but he still
called himself and his colleagues **scholastici**).
Lastly Berengario, in company with most medical men
of his time, although allowing that philosophy had
strong claims on the fundamentals of medical
thinking, nevertheless marked out a professional
territory in which medical questions could only be
answered by medical answers.

How did Berengario react to Pliny? When looking
for references to Pliny in the writings of the
academic anatomists, we should remember that part of
the function of writing anatomy books, and particularly the sort that Berengario wrote, was to
report to the reader what was to be found in the
authorities. This was the first stage of an
assessment of opinions, followed by the objections
and resolution. The use of Pliny by the anatomist
does not reflect an 'influence' of Pliny upon him,

and an uncritical presentation of Pliny´s statements does not imply acceptance until the final resolution. While Pliny is often quoted on anatomical terminology, this is only partly that the anatomist has found a preferred classical usage; it is partly also that the renaissance author is providing a guide to classical usage for his reader. Much the same is true of the early sixteenth-century works on natural history or zoology: although often seen as examples of nascent sciences struggling against the influence of the ancients, their authors were as much seeking to use nature as a guide to the authors as the authors as a guide to nature.(60) In both cases it seemed as if there was an objective reality - the structure of the body and the animated world, respectively - against which one could check Pliny´s meaning, terminology and accuracy. (Again, we see in the renaissance the beginning of the kind of practical evaluation of Pliny that appears in Rottlander´s chapter in this book.) Thus Berengario saw that Pliny had used terms for the top of the windpipe, and for the top of the gullet in man, that other authors had used interchanged. As authors sympathetic to Pliny, Berengario and Benedetti(61) distinguish, record and adopt Pliny´s usage; unsympathetic, Leoniceno(62) accused Pliny of not a mere terminological confusion, but of the gross anatomical error of confusing the two structures, that is, a windpipe that descended to the stomach and a gullet terminating in the lungs. What view was taken of Pliny, here again, depended not on the text of the **Natural History** (which was not in question), nor here directly on the evidence of the dissected corpse, but on the cultural background of the reader and, indeed, of the writer of the **Natural History**.

Conclusio totius operis

The bulk of this book consists of specialist assessments of Pliny´s **Natural History** in various subject areas. Common to them all is the admission that Pliny as collector of very many items of information did not have the expert´s understanding or precision, or indeed, purposes.

This assessement itself has a history. This chapter has described how such a history began, with Leoniceno´s attack on Pliny´s confusing some medical simples, which is parallelled by Scarborough´s critique of Pliny in the chapter on pharmacology and Morton´s on botany, which elaborate on the causes given by Leoniceno for Pliny´s errors: his misreading of his Greek sources. Secondly, it can be shown that Berengario da Carpi performed dissections of the human body in order to resolve ambiguities, locate corrupt passages and emend the text of his ancient anatomical sources, and that he used this knowledge in assessing both Pliny´s and Leoniceno´s remarks on anatomical topics. This technique of matching Pliny´s descriptions with re-creations of the physical phenomena with which he was concerned is employed by the German translation group. Here, too, archaeology offers physical objects related to the processes that Pliny describes. The temptation evident here to use something as reliable as an art-fact or a modern descriptive science as a yardstick to measure Pliny´s accuracy is also apparent in Healy´s chapter on metals and mineralogy. Any danger of an over-confident use of an approach of this sort is avoided in Greenaway's chapter on chemical tests, which are located in the context of quality-testing in a commercial world. The Hell-enists' opinion of the **Natural History** as a secondary and rather graceless text has continued to inform classical scholarship at least up until the time of Norden, as Locher allows in his chapter. In yet another area, Eastwood has described how Plinian astronomical theory was increasingly found unsatisfactory in the renaissance; a tradition of criticism of which Pedersen´s chapter is a modern example.

In a number of ways, then, this book continues an historical assessment of Pliny that began in the renaissance. The result of specialist enquiries into the **Natural History** must necessarily be as diverse as the topics to which Pliny addressed himself, and lack of a cohering argument is a com-plaint that the commentators have raised against the **Natural History.** Yet overall there is a char-acteristic Plinian approach, certainly in comparison with that of his Greek sources. Pliny´s topic is nature as the theatre of man´s actions, nature as Designed. It is not the search for natural-philosophical causes, as in Aristotle´s zoology or Theophrastus' botany, and not a categorical exercise or a catalogue of things from which to draw out inductions. Pliny´s biological knowledge is that important to man. The stars are consulted not as embodying perfection but as a guide to the agri-cultural seasons. His description of quality-testing of natural and manufactured substances, and his rejection of sophistical Greek medicine, to say nothing of his search for a simple, reliable and preferably Roman diet and **materia medica** all tell the same story. It was not just that man´s needs and concerns defined what was important in nature (see the chapters by Pedersen, Locher and Morton), but that nature was providential to the point of being divine. It was the providence of nature that led to Pliny´s physics replacing Aristotle´s in Milich´s Wittenberg, and its divinity that caused Philemon Holland to have doctrinal qualms about translating the **Natural History**. (This may serve as a reminder that the attractiveness of Pliny´s books on nature to every potential reader (and therefore their ´influence´) depended, more than simply on the utility or factual accuracy of their contents, on the existing conceptions of the reader, whether religious (Milich and Holland), cultural (Leoniceno), or academic (Berengario).

NOTES

1. In looking for progressive change in the renaissance, historians of medicine have commonly focussed on Vesalius and his attack on Galen. The result has been to pre-judge the period we are dealing with here. See for example C D O´Malley, **Andreas Vesalius of Brussels, 1514-1564,** University of California Press (the standard biography); L R Lind, **Pre-Vesalian Anatomy,** Philadelphia, 1975, and Thorndike´s account of the attack of Pliny: L Thorndike, **History of Magic and Experimental Science,** vol.4, Columbia University Press, 1934, p. 593.
2. The **Natural History** was known in the middle ages, and mss. of the eleventh century survive: H le Bonniec, **Bibliographie de l´Histoire Naturelle de Pline,** Paris, 1946, p. 23. G S Sarton, **The Appreciation of Ancient and Medieval Science during the Renaissance (1450-1600),** Philadelphia, University of Pennsylvania Press, 1955, counts fifteen incunable editions, three in the Italian vernacular. Eleven of these editions came from Venice, the work of nine different printers. Four commentaries were printed in the same period. At least 43 editions appeared during the course of the sixteenth century.
3. W Turner, **A New Herball,** London, 1551, dedication. On Turner in general, see C Raven, **English Naturalists from Neckham to Ray,** Cambridge University Press, 1947.
4. Convenient accounts can be found in Thorndike (1934), by Bylebyl, "Leoniceno", **Dictionary of Scientific Biography** (DSB) ed. C C Gi-

llespie, New York, 1970 and A Castiglioni, "The school of Ferrara and the controversy on Pliny", in E Underwood, ed., **Science, Medicine and History,** Oxford University Press, 1953, vol. 1, pp. 269-279.

5. Rome, 1492-3

6. P Collenuccio, **Pliniana Defensio Pandulfi Collenuccii Pisaurensis Iuriconsulti adversus Nicolai Leoniceni Accusationem,** Ferrara, no date (?late 1492-early 1493).

7. N Leoniceno, **De Plinii et plurium aliorum Medicorum in Medicina Erroribus,** Ferrara, 1509.

8. N Leoniceno, **De Morbo Gallico** (first ed. Venice, 1497) in **Opuscula,** Basel, 1532, pp.111r - 112r

9. Useful sources here are N Siraisi, **Arts and Sciences at Padua. The Studium of Padua before 1350,** Toronto, 1973; Q Skinner, **The Foundations of Modern Political Thought,** 2 vols., Cambridge, 1978 (I am indebted to Dr A R Cunningham for drawing my attention to this) and P O Kristeller, **Renaissance Thought and its Sources,** New York, 1979.

10. Kristeller (1979) p. 22

11. This evidence is presented in my chapter on Berengario da Carpi and anatomical commentary in A Wear, R French and I Lonie, **The Medical Renaissance of the Sixteenth Century,** Cambridge, 1985.

12. The discussion is found in the proemium to Gentile´s exposition of book 3 of the **Canon** of Avicenna. This was printed in Venice in about 1505 and again in 1522.

13. Kristeller (1979) p. 29.

14. D J Geanokoplos, **Greek Scholars in Venice,** Harvard University Press, 1962, p. 24

15. D J Geanokoplos, **Interaction of the "Sibling" Byzantine and Western Cultures in the Middle Ages and Renaissance, 1300 - 1600,** Yale University Press, p. 248

16. Geanokoplos gives the text.

17. Berengario da Carpi, **Commentaria cum amplissimis Additionibus super Anatomia Mundini,** Bologna, 1521; Gabriele de Zerbi, **Liber Anathomie Corporis Humani et singulorum Membrorum illius,** Venice, 1502. Both use the term **periti Elleni** (in various spellings).

18. Geanokoplos, "Sibling", p. 191

19. Ibid.,p.192

20. See note 11

21. For example Gentile da Foligno in his proemium to his commentary to book 3 of the **Canon** (note 21).

22. Avicenna´s natural philosophical works could be arranged in an order similar to that adopted ın the universities for the Aristotelian treatises. See the **Avicenne perhypatetici: philosophi ac medicorum facile primi opera in lucem redacta: ac nuper quantum ars niti potuit per canonicos emendata,** Venice, 1508 (Frankfurt, 1961)

23. **C Plinii Liber Secundus, de Mundi Historia cum erudito commentario V. Cl. Jacobi Milichii,** without notice of place or date; "now revised from the public lectures of the University of Wittemberg". See notes 143-5 in Eastwood´s chapter.

24. Milich (note 23) dedication.

25. R Browning, **The Byzantine Empire,** London, 1980, p. 181

26. The Hellenists commonly regarded themselves as drinking from the ´pure founts´ of Greek wisdom and eloquence, while the barbarians wallowed in ´turbid waters´, phrases used for example by Erasmus, the anatomist Sylvius and Leoniceno´s Roman pupil. On eloquence and the Hellenists´ image of themselves as a group, see Leoniceno´s letter to Politian at the beginning of the **De Erroribus,** Menochius´ letter to Leoniceno (Leoniceno´s **Opscula,** p. 55r) and Leoniceno´s reply, the letters between Leoniceno´s Roman pupil and Bonaciolus (**Opuscula** pp.146r-v) and the text of the **Antisophista** which follows, particularly pp. 147r, 170v.

27. For example Leoniceno´s pupil speaks of literary studies being pursued more by "private men" than by professional teachers. The anonymous **Antisophista** is included (p.146r) in the **Opuscula.**

28. His position is stated at the beginning of the whole Pliny controversy, in the letter to Politian.

29. **Opuscula,** p.65r

30. **Opuscula,** p.22v

31. **Opuscula,** p.39v

32. **Opuscula,** p.55r

33. **Opuscula,** pp. 55r and 61v; **De Erroribus** (ed. of 1509).

34. See for example A Benedetti, **Historia Corporis Humani; sive Anatomice,** Venice 1502, dedication; Leoniceno´s Roman pupil in his **Antisophista** (**Opuscula,** p. 146v); Benedetti to Marcus Sanutus (prefatory material to Benedetti´s **Aphorisms,** included with the **Anatomice** of 1502).

35. Very often the trio Galen, Dioscorides and Paulus (Aegineta) are Leonıceno´s preferred authors; for example **Opuscula** p.9v

36. **Opuscula,** p. lv

37. Book three of the **De Erroribus**, that is, the letter to Tottus: **Opuscula**, p. 22v

38. **Opuscula,** , pp.22v.ff

39. In the address to Politian at the beginning of **De Erroribus.**

40. **Opuscula,** pp,2r,3r,3v,5r,16v

41. Leoniceno´s **De Tribus Doctrinis** was published in the **Opuscula** of 1532 and in (Turrisanus) **Plusquam Commentum in Parvam Galeni Artem** , Venice, 1557, p. 241

42. In **Plusquam Commentum** (1557).

43. See note 6.

44. That is, Zerbi (1502).

45. See note 12.

46. The **Defensio** is unpaginated: this remark appears at sign. aii.

47. A **scholasticus** was a teacher or pupil in a **studium.** The term was used at first without a pejorative sense by the Hellenists (the Roman pupil of Leoniceno calls Montanus and his teacher **scholastici** (**Opuscula** pp. 173ff.) and Berengario and Collenuccio as academics call themselves and their colleagues by the same term. See also note 11.

48. **Opuscula,** p.44r

49. Collenuccio used the phrase twice and attributes it to Barbarus (sign. fiii).

50. Leoniceno defends himself against a charge of affectation of **Graecitas** when he introduced the Greek term **epiglossis** to correct the confusion that followed Pliny´s description of the structures of the throat.

51. **Opuscula,** p.173

52. Leoniceno, **Opuscula,** pp.29v, 43r; the Roman pupil, **Opuscula,** p. 172r

53. Leoniceno´s Roman pupil in particular attacks "the Paduan Sophist" as a **cliens** of Turrisanus, "The Monk": **Opuscula,** pp.158r, 168v. Bonaciolus makes a more general attack on monks, but both are discussing the three doctrines of Galen, and Turrisanus is intended by both. (**Opuscula,** p.156v.)

54. See note 11.

55. **Opuscula,** p.5v

56. Leoniceno´s Roman pupil reports the exchange: **Opuscula,** pp. 146v, 147r.

57. A Benedetti, ed., **C Plinii Secundi Veronensis Historiae Naturalis Libri XXXVII,** Venice, 1513. For the anatomy see note 31.

58. Leoniceno defended himself against the critics of his translations of Galen, including Berengario, in his **Contra suos Obtrectatores Apo-**

logia (published in the **Opuscula** and also separately, Venice, 1522). Falloppio considered Berengario to be the man who initiated the anatomical revolution concluded by Vesalius.

59. See note 11.

60. This purpose is seen for example in the prefatory material of the various works of Gesner, William Turner and Edward Wotton.

61. **Anatomice,** p.5r

62. **Opuscula,** p.35r

Index

Index